The Science of Musical Sound

# THE SCIENCE OF MUSICAL SOUND

John R. Pierce

**SCIENTIFIC
AMERICAN
LIBRARY**

Scientific American Books
An imprint of W. H. Freeman and Company
New York      San Francisco

**Library of Congress Cataloging in Publication Data**

Pierce, John Robinson, 1910–
  The science of musical sound.

  Bibliography: p.
  Includes index.
  1. Music—Acoustics and physics.   2. Music—
Psychology.   3. Sound.   I. Title.
ML3807.P5   1983     781′.22     82-21427
ISBN 0-7167-1508-2
ISBN 0-7167-1509-0 (pbk.)

Printed in the United States of America

Scientific American Library is published
by Scientific American Books, an imprint of
W. H. Freeman and Company, New York
and San Francisco

1 2 3 4 5 6 7 8 9 0 KP 1 0 8 9 8 7 6 5 4 3

*To Max Mathews,*

*whose Music V*

*and whose kind and patient counsel*

*started many things in many places*

# Contents

# Preface

In 1979, I was fortunate enough to receive the fifth Marconi International Fellowship, chiefly for my work on satellite communication. This had several happy effects. I met a very unusual, talented, and energetic person, Gioia Marconi Braga, through whose work and inspiration the Fellowship was brought into being, and I met her thoughtful husband, George Braga. I was able to participate in exciting events in Rome and in Sydney, Australia. Further, with the approval and encouragement of Mrs. Braga and an old friend, Dr. Walter Orr Roberts, secretary of the Fellowship Council, I was able to produce this book. I say produce, for the Fellowship supported work by Professor Jean-Claude Risset of the University of Aix-Marseille and by Dr. Elizabeth Cohen of Stanford University, both of whom have affected this book in many ways and have produced many of the sounds on the records that accompany the book. Thanks to the enthusiasm of Gerald Piel, this book has found its place among those in the new Scientific American Library.

Why was I so gratified by the opportunity to write a book on the science of musical sound? Because some thirty years ago at Bell Laboratories, all the research on speech and hearing was moved into my division. My reaction to this encounter with the science of sound was love at first sight. I knew little of this wonderful world of sound, and the man put in direct charge of the work, Edward E. David, Jr., didn't know much more.

We resolved to learn, and the learning was in itself a labor of love. A part of this learning was writing a book, *Man's World of Sound,* now out of print. We wrote alternating chapters. We learned as we wrote, and as we learned we became involved in research. We became members, and then fellows, of the Acoustical Society of America. We attended meetings, gave talks, and wrote papers.

Those were happy days that changed our lives, though many diverse things have happened since. Ed David became the President's science adviser and is now president of Exxon Research and Development Corporation. I became immersed in communication satellite work for some time, for I persuaded NASA to undertake Project Echo, and that bright balloon satellite was launched in

1960. I was closely concerned with satellite communication for some years thereafter. Nonetheless, the fascination of sound remained with me during the remainder of my years at Bell Laboratories, and after I went to Caltech as Professor of Engineering in 1971. Two of my graduate students did theses in the area of acoustics.

Early in the 1960s, my interest in sound and hearing had taken a new turn, for another colleague, Max V. Mathews, found a way to make a computer produce complex musical sounds. These sounds were at first crude, but this technique held the promise, since realized, of producing striking sounds that could not, in practice, be created by other means. Through the years, an interest in computer-produced sounds has spread to the music departments of many universities. Pierre Boulez's grand IRCAM (Institute for Research and Coordination in Acoustics and Music), associated with the Centre Pompidou in Paris, in part undertakes the study and use of computer-produced sound. Max Mathews and I worked there happily for a month in the spring of 1979, on sounds with nonharmonic partials. That work forms a part of this book.

Ever since Ed David and I together discovered for ourselves the science of sound and hearing, ever since Max Mathews and I first tinkered with computer-produced sounds, such sounds, and particularly musical sounds, have been closest to my heart. In this book I try to convey some of the fascination that I have felt.

The book describes those physical and mathematical aspects of sound waves that underlie our experience of music. Beyond that, it describes chiefly the psychoacoustics of musical sound. Psychoacoustics is the branch of experimental psychology that relates physical sounds to their perceptual features. Psychoacousticians ascertain the intensity of just-audible sounds, the least detectable differences in sound intensity, the loudness of sounds, and many more complicated aspects of sound perception.

I didn't use *psychoacoustics* in the title of this book, because I find that the term jars those who have not encountered it before. I didn't use any term denoting the science of psychology, either. As a matter of fact, many of the most important contributions to psychoacoustics have been made by scientists or engineers who weren't card-carrying psychologists. The great Hermann von Helmholtz, who in 1863 wrote *On the Sensations of Tone as a Physiological Basis for the Theory of Music,* was professor of anatomy and physiology at the University of Bonn, and later professor of physics at the University of Berlin. Wallace Clement Sabine, who founded the science of architectural acoustics early in this century, was Hollis Professor of Mathematics and Natural Philosophy at Harvard. The first systematic explorations of binaural hearing were made by a physicist, Irving Langmuir (later a Nobel Laureate), while working on the detection of submarines during World War I. Another Nobel Laureate, George von Békésy, who taught us so much about the ear, became interested in hearing as director of the Hungarian Telephone System Research Laboratory (1923–1946) and was later Senior Research Fellow in Psychophysics at Harvard. Harvey Fletcher, who, with his colleagues, first used precise and effective electronic apparatus in studies of sound and hearing, regarded himself as a physicist and served as president of the American Physical Society. Jan Schouten, a Dutch physicist, first identified the phenomenon of residue pitch or periodicity pitch, through which we hear the correct pitch of musical sounds coming from a pocket transistor radio, even though the radio is physically incapable of producing sounds of low frequency. All three Bell Laboratories men whom I have mentioned, Ed David, Max Mathews, and myself, regard themselves primarily as electrical engineers.

Why have people with such diverse and unlikely backgrounds contributed so much to a field that we must identify as a part of experimental psychology? Chiefly, I think, because progress in our understanding of sound and sensation has resulted from new modes of experimentation made possible by the clear ideas and acute tools of physics and electrical communication.

New tools and new approaches led to new discoveries. Helmholtz worked with mechanical devices: tuning forks and resonators of blown glass. Békésy and Fletcher worked with vacuum-tube devices of limited complexity. David, Mathews, and I, and our contemporaries, have had at our disposal the flexible power of the digital computer. It is true to a long tradition for me, calling myself an electrical engineer, to write about the psychoacoustics of musical sound. Electrical engineering has provided the tools—vacuum tubes, then transistors, and now digital computers—through which we can gain a deeper understanding of sound and hearing.

However, do not fear that we will become lost in complexities. Hearing is hearing. However deeply we study it, hearing remains a part of our everyday experience. The ultimate test is always how things sound. As a colleague of mine noted, a good musician is always right about sound, though the details of what he says may be wrong.

Musicians sometimes have difficulty in expressing themselves, even when they have come upon a revolutionary new idea. In striving to teach himself harmony and composition from Rameau's great *Treatise on Harmony*, Jean-Jacques Rousseau, who became a composer as well as a social philosopher and novelist, found that work "so long, so diffuse, so badly disposed" that he despaired of learning from it and sought help elsewhere. Discoverers are not always the best expositors. Rameau's revolutionary ideas are clearly and concisely expressed by his contemporaries, the distinguished French mathematician Jean Le Rond d'Alembert and the Reverend Bertrand Castel.

It is common in science for followers to organize and express matters better than originators. Furthermore, careful thought and experiment can make sense of phenomena that have been long and well used. By practicing with his sling, little David acquired a skill in the use of those laws of motion that were formulated by Galileo and Newton many centuries later.

In the field of sound and music, complicated equipment and ingenious experiments are not ends in themselves. They are means by which we can evaluate the acuity, the discrimination, the powers and limitations of hearing, a sense that we use continually, a sense through which the whole of music came into being. Electronically produced sounds should not be a part of electronics; they should be a part of the evolution of musical sound, from drum, lyre, and Stradivarius to some of today's entirely new sounds.

I write about sound as we hear it, and musical sound in particular. Of electronics, of physiology and neurophysiology, and, indeed, of sound waves and their generation, I say only what seems clearly needed for understanding and appreciating what we hear. However, relating that little from other areas of hearing and music fills a book.

I have already expressed my indebtedness to Jean-Claude Risset and Elizabeth Cohen for their part in the preparation of this book. The help of John Chowning was invaluable in the production at Stanford of satisfactory computer-generated sounds illustrating various psychoacoustic phenomena. Diana Deutsch and Max Mathews kindly allowed me to use sounds that they had produced, and Max read the manuscript of the book with care. I am grateful also to others who have read the manuscript and made helpful comments and suggestions, including Gerald Strang, Manfred Schroeder, Earl Schubert, Peter Renz, editor at W. H. Freeman and Company, and Aidan Kelly, who skillfully edited the manuscript. The contributions of Amy Malina, who found suitable illustrations, of Linda Chaput, editorial director of W. H. Freeman and Company and Scientific American Books, and of Patricia Mittelstadt, Heather Wiley, Sarah Segal, and Ellen Cash, who worked together to produce the book, were also invaluable. I am also grateful to Katherine Sipfle Taylor of the Marconi International Fellowship staff, who was my helpful correspondent for several years. I am indebted to Debbie Devoe for typing several chapters. Finally, I am grateful for the patience and encouragement of my wife, Ellen.

JOHN R. PIERCE
February, 1983

The Science of Musical Sound

# Sound and Music

**1**

The study of musical sound is important only because music is important, and because the quality of sound is important to music.

Indeed, the kind of sound that is available influences the kind of music that peoples and composers produce. Piano music is different from harpsichord music in more than the sound of the individual notes. The music of Liszt, Chopin, and Debussy exploits the unique capabilities of the piano. So do Beethoven's and Mozart's piano sonatas and concertos. But Mozart's piano was different from ours, less loud and with less dynamic range. Later music would suffer if played on it.

Composers can understand and love very early music; Edgard Varèse did. The shadings of Medieval and Renaissance vocal music are subtle. The dynamic range of early instrumental music is small. In the krummhorn, the reed was housed under a cap that separated it from the player's lips, and the range between just sounding and the loudest sound that could be produced was almost negligible. Recorders change pitch if blown very loudly (the transverse flute doesn't); so the useful dynamic range of recorders is small. Early trumpets and horns may have had greater dynamic range, but they lacked valves, and some notes could be produced only with great difficulty.

Except for the voice, musical sounds have improved greatly in range and quality in the past few centuries, and certainly up to the beginning of this century. In part, this improvement resulted from better, more easily playable instruments, especially woodwinds with better key mechanisms. In part, the improvement resulted from the increasing skill of instrumentalists and, in part, it resulted from an expansion in the range of musical sound, as composers developed and exploited new effects and new idioms.

Whether or not we wish to call such change progress, it brought an expansion in the variety of orchestral and vocal sound. Think for a moment of the sounds of Bach, Mozart, Wagner, Debussy, and Stravinsky of *The Rite of Spring*. The music that successive composers produced was different and sounded different, as if they had moved into new territory, partly in concept and organization, partly in sound itself. They did not confront the past on its own ground.

This book will say a good deal about the computer both as a means for producing musical sounds and as a means for analyzing sound. In the very early days of computer-produced sound, I felt that the computer could have a liberating effect

■ In the photograph on the preceding pages, Béla Bartók (fourth from left) records peasant songs in the village of Darazs, now in Czechoslovakia.

■ A posthorn.

■ Krummhorns.

■ A modern B-flat bass clarinet.

**GIGUE**

■ In the manuscript at the top, Arnold Schoenberg sketched out the twelve-tone row used in the published *Gigue*.

on composers, because it can in principle produce *any* sound. When I explained this to a wise musician (I think it was Milton Babbitt), he said, "It's like a grand piano in the hands of a group of savages. You know that wonderful sounds can come from it, but will they?"

Granted that something wonderful *might* come out of the computer, why should anyone bother to try? Aren't there plenty of other challenges? My impression is that there has been a need for a real, new challenge. Some composers of the late nineteenth century, and many of the early twentieth century, seemed to see no new ground to go to. After *Petrouchka*, Stravinsky turned more and more toward the past for inspiration (neoclassicism), then ended by writing twelve-tone compositions. Some British composers turned to English folk music, with pleasant rather than exciting results. Bartók did better in Hungary. Some sought inspiration in the monophonic but subtle music of other cultures. It's hard to get to the heart of what others have learned and felt from childhood. And where are you if you do?

Arnold Schoenberg, after a period of post-Wagnerian romanticism, sought to exclude the clichés of traditional harmony and structure from his compositions by his invention of the tone row and his rules for its use. In one way, the escape was a timid one. Because of the tyranny of existing instruments, he used only the twelve tones of the piano keyboard and the harmonic partials that I believe give the diatonic scale and its harmonies a strong position in any music in which notes are sounded together. Thus Schoenberg abandoned the threat of the past in a narrow, constrained, and rather academic way.

■ Gerald Strang.

Academics try to find the rules of successful past accomplishment. For example, linguists try to elucidate the structure, the use, and the history of the development of language; they don't invent new languages. J. J. Fux (1660–1741) tried to find and codify rules that would describe the practice of earlier contrapuntalists, such as Palestrina. Jean Philippe Rameau (1683–1764) tried to give a simple and compelling explanation of the harmony of his day. I think that he saw deeper than he could know, for, as I explain in Chapter 6, his ideas are closely related to the phenomenon of residue pitch, something that is inherent in our sense of hearing. Sonata form first appeared in the compositions of the Mannheim school in the mid-eighteenth century and was developed by Haydn, Mozart, and Beethoven, but not by following a set of rules slavishly. The rules and their stringency were insisted on only later.

The rules that academics deduce by studying living music have many uses. The rules are fascinating in themselves. They can be a help in listening to music. They can be a help in learning to compose music. But they do not provide a means for grinding good music out mechanically.

Lejarin A. Hiller, Jr., and his collaborator L. M. Isaacson programmed a computer to produce four-part music (which they arranged as the *Illiac Suite*) that followed most of Fux's rules for "first-species counterpoint." A little of this music sounds pleasant, but it wanders, and so does the listener's attention. When I discussed this with Milton Babbitt, he called this "specious counterpoint" and said that, anyway, the rules of counterpoint are not intended to tell you *what* to do, but what *not* to do.

I suppose the same is true of the rules of twelve-tone music, but these rules aren't derived from a study of successful past music. They are an academic guide for the production of a new kind of music. Though their purpose is escape from the pleasant old ways, this purpose can itself be escaped by an ingenious choice of the tone row. When I asked Gerald Strang, a one-time assistant of Schoenberg at the University of Southern California, what his doctoral thesis was, he said, "A symphony more in the spirit of Mozart than of Beethoven; you wouldn't know it was twelve-tone unless you were told." (Ernst Toch, rather than Schoenberg, was Strang's thesis advisor.) In some of his later computer music, Strang abandoned the twelve sacred tones entirely, and used a scale with more pitches per octave, or chose pitches without any scale.

Charlotte Moorman playing an underwater cello piece called *The Intravenous Feeding of Charlotte Moorman.*

Pierre Boulez.

After Schoenberg came not only rules, but actual prescriptions for composing music. John Cage and others chose notes at random. Pierre Boulez, Karlheinz Stockhausen, and others have proposed (and produced) totally organized music, which evolved mechanically from a choice of initial material and rules for manipulating it, or, dissatisfied with mechanical development, have written what amounted to sketches for pieces to be realized by others, using either conventional instruments or electronic means. Harry Partch invented both a new scale and an entire orchestra of new instruments to play his music. Still other composers have evoked strange sounds from conventional instruments. Perhaps the strangest was the sound of a violin burning on a New York stage, an event staged by Lamont Young and Charlotte Moorman.

While all this was going on, many talented composers produced more conventional but very attractive music, and some that managed to be individual as well. Others, like Varèse, broke new ground in a highly individual way. Nonetheless, a great deal of effort went into escape, an escape, it seems to me, into the musician's own backyard.

7

If musicians want new worlds to conquer, the computer can provide new worlds. These new worlds include all possible sounds, old and new, if only the composer has the skill and time to find and use those that will serve him. The time required may not be finite.

The first computer pieces were created by using a very primitive progenitor of Music V, a powerful sound-generating program created by Max Mathews. The very first piece, called *Pitch Variations*, was composed by Newman Guttman, a student of speech. It included primitive synthetic speech sounds. *Pitch Variations* was cleverly and subtly organized, but the organization didn't come through to the listener.

The second computer piece was mine. It was called *Stochatta* and was painfully trite, but it sounded more like conventional music, and that was what I intended. Max Mathews then composed a few pieces, including one, *Numerology,* that is both attractive and (intentionally) funny.

After we had produced a small collection of computer-played pieces, Bruce Strasser, an enterprising man at the Bell Laboratories public-relations department, had a 10-inch record made, called *Music from Mathematics*. Hoping to stir up interest among real composers, I sent copies of it to Leonard Bernstein and Aaron Copland. I received an acknowledgment from Bernstein's secretary, and the following charming reply from Copland:

■ Max V. Mathews.

■ Aaron Copland.

March 31, 1961

Dear Mr. Pierce:

It was good of you to send me the *Music from Mathematics* recording.

The implications are dizzying and, if I were 20, I would be really concerned at the variety of possibilities suggested. As it is, I plan to be an interested bystander, waiting to see what will happen next.

Thank you for your kind words about my music.

Yours cordially,
Aaron Copland

■ Edgard Varèse.

■ Milton Babbitt.

I had struck out in the first inning, but the game wasn't over. We did meet musicians who were interested, interesting, and helpful. Max Mathews and I got to know Edgard Varèse. For much of his life, Varèse had looked for "the instrument." Alas, the computer, the instrument he sought, came too late, for he was too old to start anew. Nevertheless, Varèse and his music were an inspiration. Mathews and Newman Guttman spent many hours recording and processing various sounds, including that of a buzz saw, for Varèse's *Déserts,* a composition for orchestra and tape.

We came to know Milton Babbitt, a fine man and a talented composer. Unfortunately, Babbitt was wedded to an analog synthesizer that Harry F. Olson had built at RCA. Compared with the computer, this was a dreadful, limited, ticklish machine that chewed up the paper tape that controlled it. We also came to know Otto Luening and Vladimir Ussachevsky, who, with Babbitt, were associated with the Columbia-Princeton electronic music laboratory. Ussachevsky eventually learned to use the Music V program.

Under cover, Max and I brought a promising young composer, Jim Tenney, to work on psychoacoustics at Bell Labs. He produced an interesting piece, *Noise Study,* and several others. In 1963 I gave a talk and demonstration of computer music at UCLA, arranged by Si Ramo (the *R* of TRW, as he once described himself). A group of music teachers and composers, including Gerald Strang and Ernst Krenek, sat on the platform and, at the end, commented. Strang later spent some time at Bell Laboratories. After that he produced a number of pieces by using a computer at UCLA. The talented musician and composer David Lewin visited Bell Laboratories in the early sixties, and there produced *Stage*

Otto Luening.

James Tenney.

John Chowning.

*One,* the first twelve-tone piece for the computer. Jean-Claude Risset spent two long periods at Bell Laboratories, interrupted by his military service, which he managed to spend at a French defense laboratory with a computer. Risset was a real find, a talented pianist, composer, and physicist. Also in the early sixties, John Chowning, a musician, spent a short time with Max Mathews at Bell Laboratories, and absorbed computer methods by osmosis. He went back to Stanford, infiltrated the artificial intelligence laboratory, and, turning its computer to his own use, started a marvelous organization that still endures.

During the sixties, Max Mathews and I wrote a little more computer music, but we tapered off as the computer was adopted by real musicians. Thanks to Music V, Max's last and most advanced program for generating sounds, the computer is now used to generate musical sounds at hundreds of places in the world, mostly departments of music. Many wonderful sounds and much fine music have been produced by Risset, Chowning, and others. Promising young composers now go to Stanford, to IRCAM (Boulez's Institute for Research and Coordination of Acoustics and Music, attached to the Pompidou Center in Paris), to the Massachusetts Institute of Technology, and, I am sure, to many other places that offer facilities for the computer production of musical sounds.

For a music department and for a student planning a career as a composer, computers have their pros and cons. In principle, the computer can produce any sound. In practice, it was at first hard to get the computer to produce any musically useful sound at all. All our primitive sounds were dull and monotonous, and most sounded "electronic." As will be seen in Chapter 13, good musical sounds are complicated indeed. Now that this is better understood, many attractive and useful computer-produced sounds are available, and more are found every day.

As a source of sound, the computer is much cheaper than a professional orchestra. However, getting and maintaining a really good computer installation is too expensive for most music schools. In contrast, trained bodies for a music-school orchestra are free. Furthermore, although trained singers and instrumentalists can find professional work as performers or teachers, there is a very limited musical market for experts in computer music. Perhaps it is somewhat better than the market for pianists. However, there is a nonmusical market, which is perhaps more remunerative. One computer musician from Stanford now works for the Apple computer company.

9

■ Pierre Boulez (at the left in top photograph) at IRCAM at the Pompidou Center in Paris. Rehearsal studio at IRCAM (bottom).

Suppose that a composer does have access to a good computer. What does he do with it? If he's a *good* composer, he doesn't write pieces like those that have been written for orchestra, making them louder, softer, more varied, and "bigger" than he could with conventional instruments. In principle, he could, but in my experience he doesn't.

In using new means to produce new sounds, most good composers for the computer want to make a new kind of music. They think of music as organized sound, and they want to find new organizations for new sounds. But new sounds organized in new ways will be appreciated fully only by those with new ears. When Risset blends the human voice with shushing and bell-like sounds in *Inharmonique,* the way in which he organizes these sounds is less easily grasped than forms or organization to which we have been long exposed, such as variations, rondos, or sonatas. This is true of other fine pieces by Risset, such as *Songes* and *Contours.* It is true of pieces by Chowning: *Turenas,* in which sounds whirl about through the room; *Stria,* which makes use of nonharmonic partials whose frequency ratios are the golden mean; and *Phōnē,* in which bell-like sounds change into voices. In each case, the composer has sought an organization suitable to new sounds and new effects.

Lejarin Hiller.

One sensitive musician whom I know well finds nothing but pleasing sound in some computer pieces that charm me. A lack of clear melody and rhythm makes them unmusical for her, however appealing their sonorities. Rhythm, melody, counterpoint, harmony, and various forms of repetition and transformation were the stock in trade of music in the past. We know that all this *has* worked. Will something quite different work? One can't be sure until one has made it work. Rules won't assure success. Order is order only if the listener hears it as order. He may not be able to. Even if he does, he need not be pleased. The order may seem to him monotonous or contentless. As Lejarin Hiller said, "Music is a compromise between monotony and chaos." Only musical genius can assure the wide success of computer music, if it is to succeed widely.

There are great obstacles other than the difficulty of mastering and using a new medium. For one thing, the performer is gone. Who will sit in an auditorium and listen to music coming from loudspeakers? You can't even be sure when to clap. Without some clue from player or conductor, how can you tell when the piece has ended? Some composers have resorted to pieces for voice or instruments and computer tape. This can work well, but is it the answer? Some pro-

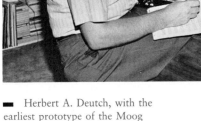

Herbert A. Deutch, with the earliest prototype of the Moog synthesizer, which he helped invent.

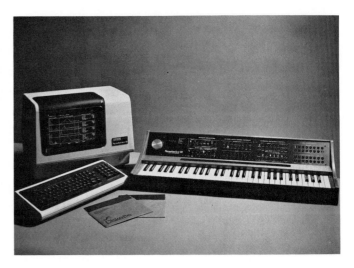

A current-model Synclavier.

pose to fiddle with controls while the piece plays, as the soulful expert of a day past fiddled with tempo and loudness while the player piano ground out its musical grist. Isn't this a little artificial?

Some composers add keyboards to microprocessors to create computerized keyboard instruments suitable for studio performance. Indeed, excellent digital keyboard instruments, such the Synclavier, are available commercially, and they have a wide range of tone and many fine features, including the recording and replaying of keyboard input. For more sophisticated effects, the Synclavier allows computerlike input for the control of sounds. Beyond this, some composers, including Pierre Boulez, have used computerlike hardware to enhance or transform the sounds produced by musicians as they play on conventional instruments.

Perhaps the digital generation of sound will lead to prodigiously popular one-man bands. In Andy Moorer's *Lions Are Growing,* which is set to a poem by Richard Brautigan, a single performer's computer-processed voice speaks, sings solo and in chords, and roars, accompanied by appropriate projected images. Moorer now works for Lucasfilm, a film-production company in San Rafael, California. Perhaps sound effects and scores for movies are the best first hope for computer-generated or computer-processed sounds.

Computer music is in an infancy of exploration and learning. It shouldn't be confined to a crib; it should get out and learn to crawl, stand, walk, run, leap, fall on its face, and be artful. Its prospects remind me of the words that the young Lincoln is said to have scratched on a slate or engraved on a wooden shovel:

Abraham Lincoln, his hand and pen;
He will be good, the Lord knows when.

LAGX1
391527 s → 15,2940230

Lions                          are                          growing
0/1        .635/5.11043        .64/5.1428                  1.18/8.63828

like            yellow            roses            on
1.19/8.703                      1,738/12.25      2.18/15,1114

the            wind            as            we
2.46/16,924    2.93/19.966    3.14/21.3256

turned            gracefully            in            the
3,465/23.429    3.85/25.9215

medieval            garden            of            their
5.02/33.495      5.44/36.214    5.72/38.026

roaring            blossoms .            Ooh,            I
6.07/40.292    6.53/43.269        7.25/47.93    8.01/52.85    8.4/54.73

want            to            turn .            Ooh,            I
8.53/56.216    8.70/57.446        9.23/60.682    10.2/67.026    10.56/69.356    10.59/69.55

on            turning .            Ooh,            I            have
last one                 11.57/75.894    12.7/83.2    12.99/85.096    13/85.15

turned .            Thank            you
13.49/88.322    14.07/92.077    14.6/95.507            15.0/98.097

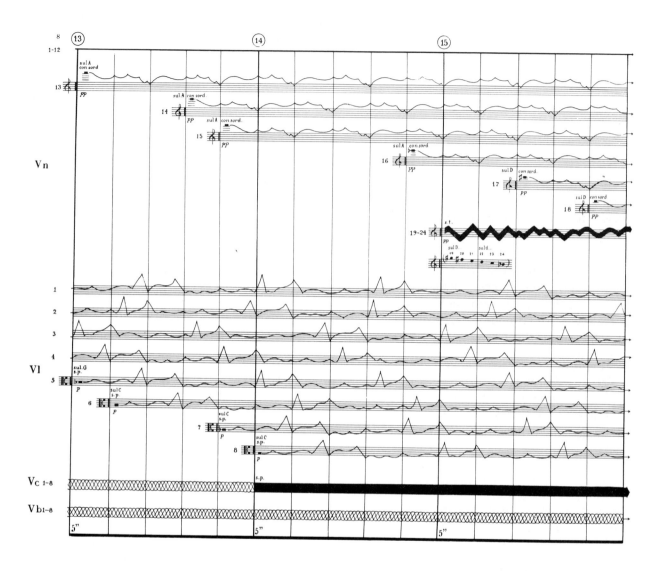

■ A page from the score of Krzysztof Penderecki's *Polymorphia*.

Recorded computer music, with or without visual accompaniments, will some-day be very popular in the home. We are already accustomed to reproducing concert-hall performances in our living rooms; why shouldn't we also learn to enjoy a kind of music that is just as varied and subtle, and, above all, is designed from the start to be heard from loudspeakers? However, such music will be accepted rather slowly if there isn't more financial support for the improvement and exploration of computer-produced sound than there has been in the past.

The reader can form some judgment of the resources of computer music and of its effectiveness from the brief recordings included with this book and a better idea from the music on the cassette recording that can be ordered from the publisher. I am sanguine about its future because sounds of all sorts fascinate me, and because the computer offers unique resources for the generation and exploration of sounds, the irreducible medium of music. Computer-produced sounds, and, indeed, electronic sounds in general, have had a profound effect on nonelectronic music. When Varèse wrote *Déserts,* for taped sound and orchestra, he was proud that he had provided such continuity of sound quality that it is hard to detect transitions from tape to orchestra. Much of Krzysztof Penderecki's music, for the conventional orchestra, has a distinct "electronic" sound quality, as does some orchestral writing of Iannis Xenakis. In such music, written at a time when computer-produced sounds were already escaping from their early electronic taint, we hear the orchestra refining sounds whose somewhat harsh qualities I myself found objectionable in early electronic music.

The influence of electronic music on some composers has been more subtle. György Ligeti, who had studied for several years with Stockhausen, abandoned the limited and difficult electronic means of tone production available at that time, but his music shows that he is acutely aware of the subtle qualities of electronic sounds and of the musical value of the sophisticated control of sound quality.

Today, many talented young composers are using the computer as an experimental tool to study the intricacies of musical sounds. Some will compose music for the computer as a musical instrument. Some will compose music for conventional instruments. All will be influenced; all will learn something new and useful. And so, I hope, will the reader of this book.

Although this book says a good deal about the computer as a means for analyzing and producing musical sounds, it is really about all kinds of musical sounds, about their well-known aspects, such as pitch, scales, consonance, harmony, and timbre, and about less-known aspects of sounds and their perception. We can't have a useful understanding of musical sounds without considering all these aspects. We will start with periodicity, pitch, and waves.

# Periodicity, Pitch, and Waves

**2**

Although wind instruments have been known for nearly five thousand years, and harps for almost as long, sounds of a definite pitch are not necessary to music. The earliest musical instruments that archeologists have found in Egypt are clappers. Perhaps song accompanied their rhythmic beat, but the use of pitch might have been limited. A knowledgable friend of mine tells me that, in very primitive music, the chief interval used is the fifth (seven semitones), though sometimes an indefinite musical third (four semitones) is also used.

Rhythm by itself can make music. In our time, Carlos Chavez composed a toccata for percussion alone. A siren is heard in Edgard Varèse's *Ionization,* but that fine piece achieves its effect chiefly by rhythm and *timbre* (sound quality).

What *is* pitch? Psychologists insist that pitch is a name for our subjective experience of periodic waveforms, rather than a physical property of the sound wave that reaches our ears. Loosely, we may use the word pitch to denote the shrillness of a sound. In this sense, the hissing sound *s* has a higher pitch than the shushing sound *sh.* In this chapter, pitch is considered to be a definite quality related to musical tones, such as those produced by violins, woodwinds, brasses, the piano, and the human voice. We hear such sounds as having a definite pitch that corresponds to particular notes of the musical scale. The present *musical* standard for "concert pitch" is that the A above middle C in the scale is 440 vibrations each second. More on pitch can be found on pages 36–37.

Musical sounds that have a definite, unambiguous pitch are *periodic.* In these sounds, something happens over and over again at a constant rate. Galileo found by accident that he could produce a sound with a pitch by scraping a brass plate with a sharp iron chisel. The tiny parallel and equidistant ridges left on the brass were a permanent witness to the vibrations of the screeching chisel that had engraved them. An old book on musical acoustics relates that Galileo produced a sound with a pitch by rubbing a knife rapidly around the edge of a milled coin. You can try this by scratching the edge of a quarter with your fingernail. The nail jerks up and down as it encounters the ridges around the edge of the coin. The sound produced does give a sense of pitch, and the faster the nail moves, the higher the pitch.

The siren provides a clear illustration of the periodicity related to musical pitch. The very siren that Varèse used in *Ionization* stood in his New York studio. When I turned the crank faster, the pitch of the sound that the siren produced

■ Sumerian seal, showing a harp (in lower right).

■ The photograph on the preceding pages shows mutation stops of an organ.

■ Brazilian samba school, an all-rhythm band.

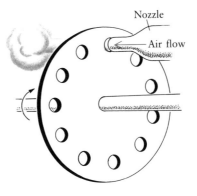

■ A simple siren. Compressed air from a nozzle passes through a circle of equally spaced holes in a rotating disk. As puffs of air pass through the holes, they excite a periodic vibration in the air. The number of puffs per second is the number of revolutions per second times the number of holes in the circle. This disk has 11 holes. If it revolves 40 times a second, it will produce 440 pulses a second, which corresponds to the pitch of A above middle C.

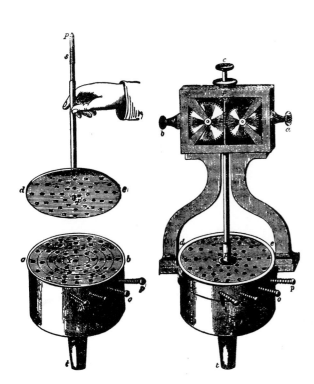

■ De la Tour's siren.

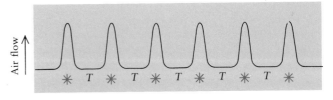

■ Periodic puffs of air produced by a siren. The time, $T$, between one puff and the next is called the period.

rose. Why was this? In order to understand, we must examine the mechanism of the siren, which was invented by Charles Cagniard de la Tour in 1819. Reduced to its essential parts, a siren consists of a rotating disk pierced by a number of equally spaced holes near its circumference, and a nozzle that directs a jet of air through the holes as they pass, as shown in the diagram on the preceding page. Such a siren emits a puff of air each time a hole passes the nozzle.

The graph on the preceding page represents the succession of puffs of air from the siren. As time passes (looking at the curve from left to right), there are intervals when no air passes through a hole (the low, flat part of the curve) and shorter intervals when puffs of air pass through a hole, shown where the curve rises and falls. The succession of puffs of air that the siren produces is periodic: that is, successive puffs are produced at equal intervals, $T$ seconds apart. The time $T$ between successive puffs is called the *period*. The number of puffs in a standard unit of time (usually one second) is the *frequency*.

The periodic pulses of air that the siren emits set up a periodic disturbance called a *sound wave* that travels through the air. When it reaches our ears, we hear this periodic wave as a sound with a definite musical pitch. When the siren produces 440 pulses per second, we hear (in concert pitch) the A above middle C. If there are 220 pulses per second, the sound is an octave lower. If there are 880 pulses per second, the sound is an octave higher. The figure on the facing page shows the frequency of various notes of the piano keyboard, together with the ranges of pitch, or *compasses,* of several other musical instruments. The piano is truly remarkable, not only for the range of frequencies that it can produce, but for its range of loudness and for the comparative ease with which a player can produce tones. How violinists struggle with fingering and to play on pitch! How trumpeters purse their lips and blow!

Only sources of sound that are periodic have a clear, definite, unarguable pitch. In this chapter, and in several that follow, only periodic sounds and their pitches are considered, for such sounds form the basis of scales and traditional harmony.

We might think that the relation between frequency and musical pitch was discovered by means of the siren. The number of pulses of air that the siren produces in a second is the number of holes around the disk times the number of times that the disk revolves each second. We can drive the disk at a high speed

■ The pitch of periodic musical sounds is determined by their frequency; that is, by the number of periods per second. This diagram relates the notes of the musical scale, the positions of the keys of the piano, and the ranges of various musical instruments to the corresponding frequencies. What about periodic sounds having frequencies much above or below the range of the piano keyboard? For such sounds, changes in frequency don't correspond to clear musical intervals, though the sensation of pitch does go up and down with frequency. Such sounds don't have a clear or useful *musical* pitch.

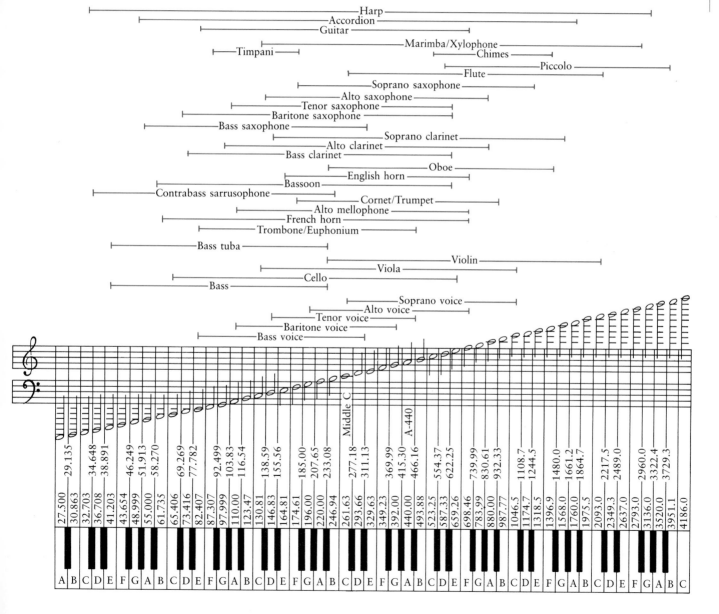

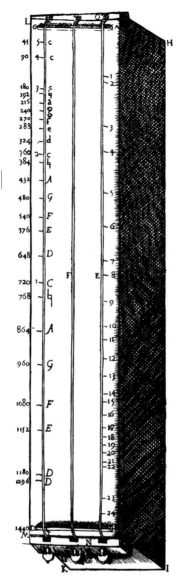

The monochord, used to relate relative lengths to pitch, as depicted in Mersenne's *Harmonie universelle*.

22

by turning a crank attached to a train of gears. If we measure the number of times per second that we turn the crank, we can calculate the number of times per second that the disk rotates, and hence the frequency of the pulses at each pitch that the siren produces.

In fact, the relation between frequency and pitch was discovered much less directly, long before the siren was invented. In *Dialogues Concerning Two New Sciences,* published in 1636, Galileo clearly explained the relation between pitch and the frequency of vibration of a string, but he wrote only of the relative numbers of vibrations per second corresponding to various musical intervals.

In *Harmonie Universelle,* also published in 1636, the French cleric, philosopher, and mathematician Marin Mersenne relates pitch to the actual number of vibrations per second. Like Galileo, with whose work he was familiar, Mersenne knew how the frequency of vibration varies with the length of a stretched string (frequency is proportional to the reciprocal of the length*), with tension (it is proportional to the square root of the tension), and with mass per unit length (it is proportional to the reciprocal of the square root of mass per unit length). Putting all this together, we find

$$\text{frequency} = k \times \frac{\sqrt{\text{tension}}}{\text{length} \times \sqrt{\text{mass per unit length}}}.$$

But by what factor $k$ must we multiply this product of quantities in order to get the actual number of vibrations per second? Mersenne found the correct factor by counting the number of vibrations per second of long strings, including a hemp cord 90 feet long and $1/12$ inch in diameter, and a brass wire 138 feet long and $1/48$ inch in diameter.

Not everyone immediately appreciated or accepted these ideas. According to Samuel Pepys's diary entry for August 8, 1666, Robert Hooke told him that "he is able to tell how many strokes a fly makes with her wings (those that hum in their flying) by the note that it answers to in musique during their flying." This Pepys characterized as "a little too much refined."

---

*For any number *n*, the reciprocal is 1/*n*.

■ Greek vase painting of citharist.

The pitch of musical sounds that we now know to be periodic was a central aspect of music long before frequency had been related clearly to pitch by Galileo and Mersenne. Many early peoples used pitch in an orderly and effective way. Like other peoples, the Greeks must have noticed from the earliest times that plucked strings vibrate. Various Greek philosophers associated high notes with swift motion and low notes with slow motion, though they gave no exact relation of motion to pitch. The new discovery that the Greeks made and passed on to posterity is the wonderful numerical relation between the lengths of strings and musical intervals. This discovery is commonly attributed to Pythagoras.

Imagine a stretched wire of length $L$, as shown at the top of the first illustration on the next page. If the wire is plucked, it will emit a sound with a definite pitch, say, middle C. If the tension of the wire is kept constant, but the length is shortened by placing a solid wedge somewhere along the wire, the pitch of the plucked wire is higher. The figure shows the relation between the lengths of the plucked wire and the pitches produced. For example, if the wire is shortened to two-thirds of its original length, it will sound the note that is a *fifth* (seven semitones) above the original note.

The Greeks had a rather mystical regard for number and proportion. They were gratified to find a relation between the ratios of the whole numbers and familiar musical intervals. They sought similar harmonious relations for the proportion of buildings. This was characteristic of their thought. Plato, for example, identified the five regular polyhedra (see the next page) with the four elements and the universe (tetrahedron, fire; cube, earth; octahedron, air; icosahedron, water; dodecahedron, the universe).

To the modern mind the relation between the lengths of strings and musical intervals is empirical, and we seek some physical explanation for it. It is in the spirit of science to try to get behind the regularities of complex phenomena and find their explanation in simple terms. Today we know that the periodic nature of musical sounds arises from the nature of waves, in air and in strings.

■ A stretched string vibrates with a particular frequency and gives a tone of a particular pitch. Here we assume this pitch to be middle C. If we keep the tension the same, but use a wedge to reduce the vibrating length to 5/6 of the original length, the frequency increases by 6/5, and the pitch goes up a minor third, to E♭. Other fractional reductions in the length of the string result in other musical intervals: 4/5, major third, E; 3/4, fourth, F; 2/3, fifth, G; 1/2, octave, C.

| Length | Pitched length shown | Interval above original length | |
|---|---|---|---|
| | | Name | Number of semitones |
| ├── $L$ ──┤   Stretched wire | C | — | — |
| ├── $(5/6)L$ ──┤ | E♭ | Minor third | 3 |
| ├── $(4/5)L$ ──┤ | E | Major third | 4 |
| ├── $(3/4)L$ ──┤ | F | Fourth | 5 |
| ├── $(2/3)L$ ──┤ | G | Fifth | 7 |
| ├── $(1/2)L$ ──┤ | C′ | Octave | 12 |

■ The five regular polyhedra.

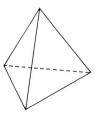

Tetrahedron

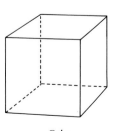

Cube

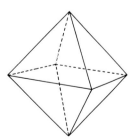

Octahedron

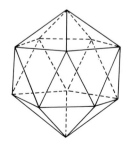

Icosahedron

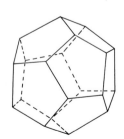

Dodecahedron

■   Ripples on water.

We have all seen the circles of ripples that move outward when a raindrop falls into a quiet pool, or when we drop a pebble into smooth water. In a similar way, a disturbance of the air moves out from the disturbing source. However, sound waves do not travel on a surface, but through the air in all directions. The air in a sound wave does not move bodily, as water flows in a stream, but only locally. One part of the air imparts motion to that ahead, as might happen if, in a long line of closely spaced people, a person at the end gave a push to the one ahead, and that one, in turn, pushed the next. We can imagine a disturbance traveling to the head of the line without anyone taking a step forward.

We all experience sound in air, but we can't see sound waves. Furthermore, they pulsate so rapidly that we can't feel their individual pulsations, except, perhaps, in the lowest notes of a pipe organ. We can visibly represent the motion of the air in a sound wave with a contrivance made of a series of little weights connected by springs (see the figure below). If we give a sharp blow to one end of this device, the weights down the line move forward and back in turn, and we see a wave travel along the string of weights and springs. The figure shows successive positions of the masses, as a single wave, or pulse, travels to the right.

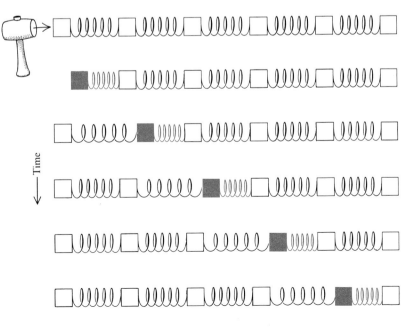

■   A wave can travel along a sequence of weights connected by springs. Waves travel through air in a similar manner, for air has both mass and springiness, or elasticity.

 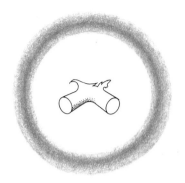

Time ⟶

26

A single sound wave excited by an exploding firecracker is an expanding spherical shell of compression in the air.

This contrivance produces waves that, unlike sound waves, travel in only one dimension, a straight line. However, the springs and weights accurately represent the two properties of air that allow the propagation of sound waves—elasticity and mass.

We are familiar with the elasticity of air from experience with automobile tires and bicycle pumps. When we press on air, it gets smaller and fits into a smaller volume. Conversely, air under pressure expands if it can, as we see when we blow into a toy balloon. That air has mass is evident in every breeze and wind strong enough to make visible objects move. Sailboats and the leaves of trees move in the wind because the moving air imparts to them some of its momentum, a property of moving things that have mass.

A single sudden disturbance, such as the explosion of a firecracker, pushes the air next to the disturbing object. Because air has mass and elasticity, it resists and is compressed. The compressed air then expands again, pushing in all directions against the air around it. The surrounding air in turn becomes compressed, forming a shell of compressed air a little distance from the original disturbance. The expansion of the air in this shell creates yet another shell, farther out, and so on (see the above illustration). Thus, we may think of a single sound wave as an expanding shell of compression. Successive layers of air are compressed and decompressed as the wave moves outward from the source of disturbance, but each individual air molecule moves only a little distance out and back. A steadily vibrating object, such as a plucked string, starts an expanding shell of compression with each vibration. Thus, we can visualize the emitted tone as a series of expanding shells (see the left-hand illustration on the facing page).

In the figure on page 25 and in actual sound waves, the movement of the medium, whether springs and weights or air molecules, is back and forth along the direction in which the wave itself travels. Such waves are *longitudinal*. In contrast, waves in water are *transverse*: that is, most of the motion of the water is up and down, *sideways* to the direction of the visible wave. Plucked strings also exhibit transverse waves.

The tension in a stretched string is a force that tends to straighten out bends in the string. Because of the tension, a stretched string, when pulled aside at one

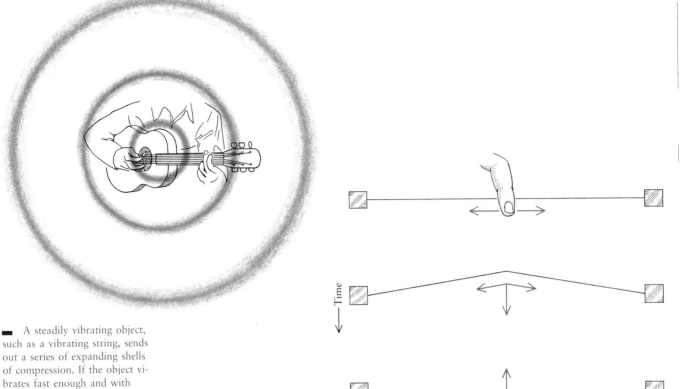

A steadily vibrating object, such as a vibrating string, sends out a series of expanding shells of compression. If the object vibrates fast enough and with enough force, we can hear the succession of shells as a tone.

The force of tension (colored arrows) and the inertia of a plucked string keep the string vibrating back and forth.

point, pulls back. However, because it has mass and, thus, inertia, it does not simply return to a straight line after it has been released, but moves on, to bend on the opposite side. Here the force of tension acts in the opposite direction and pulls the string back again. Thus the tension and mass of the string act together to keep it vibrating from side to side (see the right-hand figure above).

We can look at the way in which vibrating strings move from side to side in another way; we can consider the vibrations to consist of waves that travel along the string and are reflected at its ends. If we displace and release a string, we set up a wave that travels down one side of the string to the end, where it is reflected as a sort of echo; then it starts back along the string in the other direction and as a bend on the other side (see the figure on the next page).

Under favorable conditions, waves in strings can be both seen and felt. Inside the glass facade of the New York State Theater at Lincoln Center hangs a sort of curtain made of separate strands of metal beads strung on wires or cords. During an intermission, I was standing on a walkway near the very top of these strings of beads, and I idly plucked a string to see how a wave would travel down it. A noticeable transverse wave did indeed travel down the string of beads.

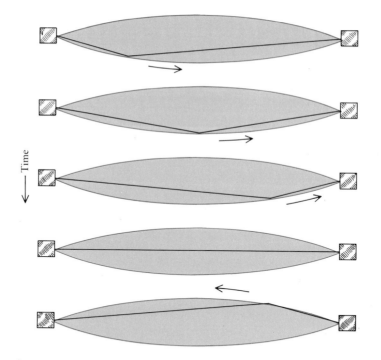

Time

28 | ◼ A plucked string appears to widen into a curved ribbon whose curved edges narrow to a point at each end. However, the actual shape of the string at any moment is a straight line sharply bent at one point. These "snapshots" of the bend show it traveling as a transverse wave along the stretched string. In the last snapshot, the wave has been reflected at the right end of the string, and the bend is moving leftward on the opposite side of the string.

I wondered if I could observe such a wave at home. First, I fastened a heavy thread to a door knob and unwound several tens of feet. Then I plucked the string. The results were inconclusive: I couldn't be sure that I had observed waves, and the thread was a terrible mess when I had rewound it on the spool. I had better luck outside, with a nylon fishline. After tying one end to a tree, I unwound about sixty feet of line, stretched it lightly, and plucked the end that I held. I couldn't convince myself that I could see a wave, but I could certainly feel it as a series of jerks or pulses. The wave that I had set up traveled to the tree, was reflected there as an echo, was reflected again at the spool that I held, and traveled back and forth several times before dying out. I felt a little jerk each time the wave reached me.

Whether waves travel as longitudinal waves in air, or as transverse waves along stretched strings, or in water, they are reflected when they encounter a solid, immovable obstacle. (The reflection of waves is discussed in Appendix E.) A reflected wave is called an echo. How long does it take an echo to return? That depends on the distance and on the speed of the wave. In a stretched string, the velocity, $v$, of the waves, measured in meters per second, is given by

$$v = \sqrt{\frac{T}{M}}.$$

In this equation, $T$ is the tension of the string, measured in newtons, and $M$ is the mass per unit length of the string, measured in kilograms per meter. (Why this is so is explained in Appendix D.) Thus, movement of the waves is slower in a more massive string and faster in a string that is stretched more tightly.

A violin dealer once came to Max Mathews for advice about choosing strings for a violin. He knew that a steel string required more tension to produce the same note as a gut string, but he didn't know how much more, and he feared that trying the steel strings might damage the violin. Mathews told him to estimate the increase in tension by weighing the strings. If a steel string is more massive than a gut string, its tension must be increased in the same ratio to give the same frequency and pitch.

The velocity of sound waves is not affected by changes in air pressure, because compressed air has both greater density, which slows down the waves, and greater elasticity (resistance to compression), which speeds up the waves. However, the velocity of sound waves does increase with temperature, because the elasticity of air results from the motion of its molecules, which move faster as the temperature is increased.

At room temperature (conventionally, 20 degrees Celsius or 68 degrees Fahrenheit), the velocity of sound is 344 meters per second, 1,128 feet per second, or 769 miles per hour. This velocity was first measured, somewhat inaccurately, by Mersenne in about 1636. He measured the time interval between seeing the flash of a gun at a known distance and hearing the sound. A more satisfactory measurement was made later, in about 1750, under the direction of the Academy of Sciences in Paris. Knowing the velocity of sound in air, if we count the seconds between seeing a flash of lightning and hearing the associated thunder, we can easily tell how far away the lightning struck. The distance in feet is about 1,000 times the number of seconds—about five seconds for a mile, or three seconds for a kilometer.

Under certain circumstances, echoes demonstrate the relation between frequency and pitch. Perhaps the simplest way to hear this is to stand in front of a receding sequence of regularly spaced vertical surfaces. These could be a sequence of solid seats, as in a stadium, or the seats of a Greek theater, or even a long flight of broad concrete steps. A handclap or any other sharp sound will be reflected from successively more distant surfaces, so that it returns as a series of echoes. What is the time interval between successive echoes? Suppose that the reflecting surfaces are spaced W apart. In reflection from the next farther sur-

Greek theater at the University of California, Berkeley.

A single handclap in front of a receding sequence of steps or seats, as in a football stadium or a Greek theater, produces a periodic succession of echoes. If they are loud enough, they can be heard as a sound having a definite pitch.

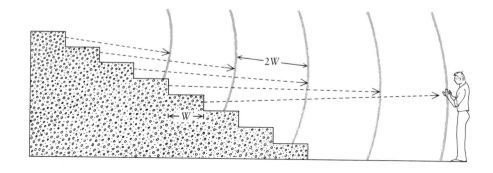

face, the sound has to travel an added distance W in reaching the surface and an added distance W in getting back (see the above figure). For an observer at the sound source, the time interval, T, between one echo and the next is two times W divided by the velocity, $v$, of sound:

$$T = \frac{2W}{v}.$$

The frequency, which is the number, $f$, of echoes per second, is thus

$$f = \frac{1}{T} = \frac{v}{2W}.$$

■ The overhang at the University of California at Santa Cruz.

■ The echo from a handclap under a high overhang must travel up to the overhang and back, a distance approximately $2H$, every time it returns to the observer. The echo from a handclap in the middle of a hallway must travel to the wall and back, a distance $W$, every time it returns.

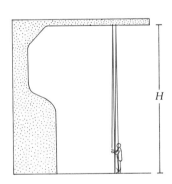
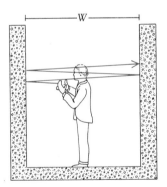

If the distance, $W$, between seats were 3 feet, the frequency would be 188 echoes per second. This corresponds to a pitch of about the first G below middle C. For steps having a standard depth of 10 inches, the frequency would be 677 echoes per second, or a pitch between those of the second E and F above middle C.

It is easier to hear successive echoes from two hard, parallel surfaces. Recently, I succeeded in hearing a sequence of echoes as I stood on a concrete walkway under a projecting canopy of the Drama Building of the University of California at Santa Cruz. When I clapped my hands, I heard a distinct sequence of claps, or pulses, of decreasing intensity. The overhang was about twenty feet above the concrete on which I stood; so the reflections repeated with a frequency of about 28 per second. I didn't hear them as a pitch, but as a sequence of sharp sounds.

The hall of our house is four feet wide; for someone standing in the middle of the hall, the reflections coming alternately from *both* walls would have a frequency of

$$f = \frac{v}{W} = \frac{1128}{4}.$$

This calculation gives a frequency of 282 echoes per second—roughly, the C♯ above middle C. I can't say that I actually hear this pitch when I clap my hands in that hall, but on occasion I have heard a tonelike flutter when I clapped my hands in some rooms and halls that had hard, uninterrupted parallel walls.

With each vibration of a stretched string, a transverse wave travels the length of the string twice—once each way. Thus the frequency, the number of vibrations per second, is given by

$$f = \frac{v}{2L}.$$

Here $L$ is the length of the freely vibrating string, and $v$ is the velocity with which a transverse wave travels along the string. It has already been noted that this velocity is the square root of the ratio of the tension, $T$, to the mass per unit length, $M$. Thus, the frequency of vibration of the stretched string of a musical instrument is

$$f = \frac{\sqrt{T/M}}{2L}.$$

All stringed instruments are tuned by adjusting the tension, $T$, of the strings. Thus, tightening a guitar string raises the pitch of the note that it produces, because an increase in $T$ causes an increase in $f$. While playing the guitar, however, the musician continually changes the pitch of the strings by changing their length, $L$. When a string is held down on one side of a fret, the vibration of the string on that side is suppressed; the rest of the string, being shorter than the whole, then produces a higher note when it is plucked. Although the main change in pitch is caused by the reduction in length, the guitarist can also introduce minor variations into the sound of a vibrating string by wiggling the finger

Guitarist.

that holds the string down. This causes very slight fluctuations in $T$, and thus in $f$, so that the string does not produce a steady tone, but a vibrato.

No convenient way has been devised to "tune" the mass of a string. Thus, it must be chosen carefully so that, at a reasonable tension, the string will produce a sound of the desired pitch and loudness. In a piano, the bass strings are heavily overwound to increase the mass, and they are longer than the treble strings. Because of the great mass, a wave travels along the strings slowly; because of the long length, it has to travel far. The massive bass strings give a loud sound. In order to make higher notes comparably loud, two strings (also overwound) are used per note in the midrange and three strings (plain) in the upper range.

In wind instruments, the frequency is determined by the time it takes a sound wave in air to travel from one end of a tube to another. A pipe of length $L$ that is open at both ends vibrates at a frequency

$$f = \frac{v}{2L}.$$

A pipe closed at one end gives a sound one octave lower than an open pipe of the same length. In an organ, for example, an open 8-foot pipe should have a frequency of about 70 vibrations per second, roughly two octaves below middle C. A 16-foot open pipe or an 8-foot pipe closed at one end would sound another octave lower.

Organ pipes are arranged in *stops*, or ranks of pipes. A stop contains one pipe for each key of the keyboard, and can be referred to by the length of its longest pipe. Thus all the pipes of a 16-foot stop are twice as long as the corresponding pipes of an 8-foot stop, and all their pitches are an octave lower.

The pipes of different stops differ not only in length but also in shape and composition, which gives them different *timbres,* or qualities of sound. Furthermore, two or more stops can be played simultaneously to give special timbres. *Mutation stops* are designed for just this purpose: the nazard, or 2⅔-foot stop, gives sounds pitched an octave and a fifth above those of an 8-foot stop; the tierce, or 1⅗-foot stop, gives sounds pitched two octaves and a major third above those of an 8-foot stop; the larigot, or 1⅓-foot stop, gives sounds pitched two octaves and a fifth above those of an 8-foot stop. By playing these stops

33

■ Piano strings.

together with 8-, 4-, or 2-foot stops, organists produce sounds having strange, juicy timbres that may sound to the uninitiated like a lot of wrong notes going along with the melody.

We have seen in this chapter that the pitch of a sound depends on the frequency of the vibrations that produce it. In a siren, the frequency, and hence the pitch, of the sound depends on the number of holes that pass the nozzle each second. Echoes from a sequence of equally spaced obstacles can produce a tone whose frequency is determined by the number of echoes that reach the observer each second. In a vibrating string or a sounding organ pipe, the pitch is determined by the time it takes a transverse wave or a sound wave to travel back and forth along the string or pipe. The next chapter explores ideas of frequency and periodicity in more detail.

# A Note on **Musical Pitch**

"Pitch" originated as a musical term and has become a psychological term used to designate a perceived quality of sound.

In music, the pitch of musical sounds was perceived long before the physical basis for pitch was understood. One of the great musical (and psychological) discoveries is that for periodic musical sounds, such as those produced by the organ, strings, winds, and the human voice, pitch is tied unalterably to the periodicity or frequency with which the waveform of the sound repeats.*

Periodic musical sounds are made up of many harmonically related frequency components, or *partials*, of frequencies $f_0$, $2f_0$, $3f_0$, $4f_0$, and so forth. Such sounds have many perceived qualities besides pitch. One of these other qualities is shrillness, or brightness. A sound with intense high-frequency partials is bright, or shrill. A sound in which low-frequency partials predominate is not bright, but dull.

When you listen to periodic musical sounds on a hi-fi system, you can change the brightness by turning the tone control, but this doesn't change the pitch. The brightness depends on the relative intensities of partials of various frequencies. Turning the tone control can change the relations of the partials, but won't change the periodicity of the sound, which is the same as the *fundamental*, the frequency of the first partial, $f_0$.†

Sounds that are not periodic musical sounds are not as clear and distinguishable in pitch and brightness, but some of them can be granted pitch by a sort of musical courtesy.

---

* There can be an ambiguity of an octave in the pitch of a periodic musical sound, because musicians sometimes (though rarely) report the pitch as being an octave away from where it falls according to its periodicity.

† The first partial need not be physically present in the sound wave, but its absence doesn't alter the pitch. This psychological phenomenon will be discussed in Chapter 6.

Among these are sine waves (pure tones), the tones of bells, the clucking sound that we can make with the tongue and the roof of the mouth, the somewhat related sound of the Jew's harp, and the sound of a band of noise.

Sine waves are peculiar in that they consist of a single harmonic partial. The sense of pitch that they give is not as certain as that of other periodic sounds; it can differ a little with intensity, and between the two ears. For other periodic sounds, the sense of the octave is very strong, for the partials of a sound $a'$ (that is, an octave above sound $a$) are all present in sound $a$.* The sense of the octave is not strong with sine waves. Furthermore, because sine waves contain only one frequency component, their brightness is tied inextricably to their pitch.

Musically trained people react to sine waves and their pitches much as they react to periodic musical sounds. Naive people may react differently. By asking naive subjects to relate frequency changes of sine waves to a doubling of pitch, psychologists found a *mel scale* of pitch (for sine waves). In the mel scale there is no simple relation between frequency and pitch; nothing like the octave shows up. I think the mel scale is a scale of brightness, not of pitch. It might be possible to check this by using musical sounds whose brightness and pitch could be varied independently.

The sounds of orchestral bells and of tuned bells are not periodic, and these sounds do not have all the properties of periodic musical sounds. One can play tunes with bells, and the pitches that are assigned to bells can be explained largely in terms of the frequencies of prominent, almost-harmonic partials.

Clucking sounds and shushing sounds (bands of noise) have a brightness, but no periodicity. Oddly, *we can play a recognizable tune with these sounds,* even though they cannot be heard as combining into chords or harmony. Apparently, in the absence of a clear pitch, brightness can suggest pitch. This seems natural. When we play a scale on a musical instrument, the brightness increases as we go up the scale. But the "pitch" of clucks or bands of noise is only a *suggestion* of pitch. It depends on the frequency at which the brightness spectrum peaks, and this (and therefore the "pitch") changes when we turn the tone control.

Even periodic sounds can be constructed to give queer pitch effects, but such sounds are *not* produced by musical instruments. For periodic musical sounds, the pitch is tied firmly to their periodicity, the frequency of the first harmonic partial. The only mistake we can make in "confusing" pitch, a sensation, with periodicity, the numerical frequency of the fundamental, is that of offending psychologists.

*If a sound has frequency $a$, its partials have frequencies $2a$, $3a$, $4a$, and so forth. The octave of $a$ has frequency $2a$, which has partials $4a$, $6a$, $8a$, and so forth. All frequencies in this latter series also occur in the first series.

# Sine Waves and Resonance

We have seen that sound waves travel from the source of sound to our ears as fluctuations in the pressure of the air. Different sorts of fluctuations cause us to hear a wonderful variety of different and interesting sounds, sounds that we can identify and appreciate. Among these are musical sounds, in which the air pressure rises and falls almost periodically with time and which have a pitch that corresponds to the frequency of this nearly periodic rise and fall in air pressure.

We can pick up the variation of sound pressure with a microphone, and we can use a cathode-ray oscilloscope to trace out the way that the sound pressure varies with time. Thus we can see the waveform of a sound wave. Is there a method that allows us to analyze sound waves and understand why the ear responds differently to different sounds? There is, indeed, and that is the subject of this chapter.

Among the sounds used in the laboratory there is one, called a *pure tone,* or *sine wave,* in which the air pressure rises and falls sinusoidally with time, as shown in the figure on the facing page. A mathematical explanation of sine waves is given on pages 58–59.

Why are pure tones, with their sinusoidal variation of air pressure, important for understanding musical sound? We almost never hear pure tones, except in the laboratory, or when we listen to a tuning fork that isn't struck too hard. Pure tones sound very uninteresting. When the pitch is low, they sound like the hum of a malfunctioning radio. Pure tones of higher pitch are steady but not bright or interesting whistles. Pure tones are also, in many ways, odd and unnatural sounds. In a reverberant room we can't sense the direction from which they come. Some people hear the pitch of a pure tone differently in the left and right ears; this is called *diplacusis binauralis.* Even for people with normal hearing, pure tones change pitch noticeably when they are made very loud; this effect is nearly absent for musical sounds. Finally, in the low range of pitch, below middle C, pure tones aren't nearly as loud as instrumental sounds of the same power and pitch.

Happily, we encounter pure tones chiefly in musical and psychological books and in acoustical laboratories, and sometimes as the output of inferior or ill-used electronic music synthesizers. Why do pure tones have any place in experiments on hearing and in the understanding of musical sounds? The reasons are partly mathematical, partly physiological and perceptual.

A sine wave or pure tone.

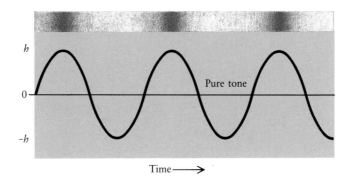

A pure tone is simply a tone for which the air pressure varies sinusoidally with time; that is, as a sine wave. A sine wave is a mathematical function that has unique and important properties. Mathematically, we can represent any periodic variation of air pressure with time as a sum of sinusoidal components (as we will see in looking at the illustrations on page 43). Furthermore, in responding to various sounds, the mechanism of the ear partly sorts out sinusoidal components of different ranges of frequency. When we listen to a musical sound, such as that of a voice or of a clarinet, different nerve fibers that go from the ear to the brain are excited by different ranges of the sinusoidal components that we can find by mathematical analysis of the sound wave.

The shape of a sine wave is completely described by three attributes: *amplitude, period,* and *phase.* First, the *amplitude* represents the maximum displacement of the varying quantity from its average value. In the illustration above, the amplitude is $h$. In a sound wave, air pressure periodically rises to a high pressure ($P$) and falls to a low pressure ($-P$), above and below the average air pressure.

Second, the *period, T,* of a sine wave is the time between amplitude peaks, usually measured in seconds. The reciprocal of $T$, $1/T$, therefore gives the number of peaks per second, which is the *frequency, f,* of the sine wave:

$$f = 1/T.$$

42   ■  A sine wave is characterized completely by three quantities: its *amplitude* or extreme height, its *period* (the time between one peak and the next), and its *phase*, which we can take as the time when the wave crosses the axis going upward. In this figure, the solid curve crosses the axis going upward at time $t = 0$. The dashed curve peaks at $t = 0$ and crossed the axis going upward at a time $T/4$ earlier (at $t = -T/4$, in which $T$ is the period of the wave). The solid curve is called a *sine* curve, and the dashed curve is called a *cosine* curve. A sine curve and a cosine curve differ in phase only.

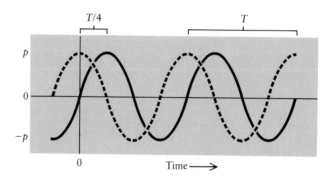

Frequencies used to be stated as cycles per second, or cycles for short. Today the term *Hertz,* honoring the physicist Heinrich Hertz (1857–1894), is used to designate cycles per second. The abbreviation of Hertz is Hz.* For example, a sine wave with a frequency of 440 Hz has a period given by

$$T = 1/440 = .0022727 \text{ seconds.}$$

Similarly, a sine wave with a period of 1/1000 or 0.001 second will have a frequency of

$$f = 1/0.001 = 1,000 \text{ Hz.}$$

Third, a sine wave has a *phase*. The two sine waves shown in the figure above have the same frequency and amplitude, but have different phases because they reach their amplitude peaks and cross the horizontal axis at different times.

The French mathematician François Marie Charles Fourier (1772–1837) invented a type of mathematical analysis by which it can be proved that any periodic wave can be represented as the sum of sine waves having the appropriate amplitude, frequency, and phase. Furthermore, the frequencies of the component waves are related in a simple way: they are all whole-number multiples of a single frequency, $f_0$, $2f_0$, $3f_0$, and so on.

---

* Often people use *frequency* to describe the rate of oscillation of complex waveforms that are made up of many sinusoidal components of different frequencies. In this book I try to use *frequency* and *Hertz* only for sine waves, and *periodicity* for the number of cycles per second of complex waveforms. This may seem awkward, but it is unambiguous.

■ A square wave.

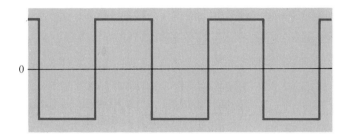

A Fourier representation of a complicated wave can require very many components, even an infinite number. However, to approximate a wave to a desired
degree of accuracy, fewer components may be adequate. A deceptively simple-
looking wave whose Fourier representation requires an infinite number of components is the *square wave* (see the figure above). A square wave having amplitude 1 and frequency $f_0$ can be represented as the sum of sine waves having
frequencies $f_0$, $3f_0$, $5f_0$, $7f_0$, (and so on, indefinitely), amplitudes 1, 1/3, 1/5, 1/7
(and so on, indefinitely), and the proper phases (see the figure below). (If the
number of components being added is finite, as it will be in any real amplifier,
the resulting square wave will not have exactly sharp corners; that is, it will be
somewhat distorted.) We should notice that a square wave is peculiar in being
made up of only odd frequency components. Most periodic sound waves consist
of both odd and even frequency components (although closed organ pipes and
some wind instruments do have predominantly odd frequency components).

■ **A.** Three waves, of frequencies $f_0$, $3f_0$, and $5f_0$.
**B.** The wave that results from
adding together the three waves
in part **A.**
**C.** The approximation to a
square wave produced by adding
together the first 19 components
in the same series as in part **A.**

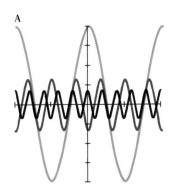

A

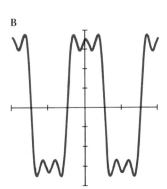

B

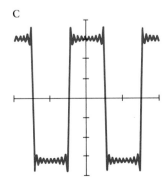

C

The first twelve natural harmonics of $C_2$.

There are three different systems for naming the sinusoidal frequency components of a periodic sound, as shown in the table below. Notice that the number of the harmonic or partial is the same as that of the relative frequency; for example, $5f_0$ is the fifth harmonic or the fifth partial. However, $5f_0$, the *fifth* harmonic, is the *fourth* overtone; so it is important not to confuse these three terms.

Systems for naming frequency components.

| Frequency | Harmonics | Overtones | Partials |
| --- | --- | --- | --- |
| $f_0$ | Fundamental | Fundamental | First partial |
| $2f_0$ | Second harmonic | First overtone | Second partial |
| $3f_0$ | Third harmonic | Second overtone | Third partial |
| $4f_0$ | Fourth harmonic | Third overtone | Fourth partial |

It is convenient to use the term harmonic to deal with strictly periodic sounds. However, some sounds (usually percussion) used in music consist of frequencies that are not harmonic; that is, are not integral multiples of the lowest frequency. For example, the frequencies of a "free" (lightly supported) vibrating rod or bar might be

$$f_0, 2.756f_0, 5.404f_0, 8.933f_0, \text{etc.}$$

It would sound rather illogical to call these higher frequencies "nonharmonic harmonics"; so instead they are called "nonharmonic partials." It is also convenient that the numbering of the partials agrees with the numerical order of the frequency. The lowest frequency is always the first partial; the next higher frequency is the second partial; and so on.

In principle, a sum of sine waves can be used to represent any sound. If the sound is of finite duration, as actual musical sounds are, we must use an infinite number of different harmonics to represent the sound by means of a *Fourier*

■ Hermann von Helmholtz.

■ Helmholtz resonator.

*integral* (which is the function that sums all the harmonics into the sound wave). However, trying to represent actual sounds as sums of *true* sine waves, which persist from the infinite past to the infinite future, is a mathematical artifice. Consider the (nearly) periodic sounds produced by musical instruments. A sum of harmonically related sine waves doesn't correctly represent such a sound, because the sound starts, persists a while, and dies away.

45

Let us also consider "noisy" sounds, such as the hiss of escaping air or the *sh* or *s* sounds of speech. Such sounds can be represented as the sum of sine waves that have slightly different frequencies. If you make the *sh* twice, the wave forms won't look exactly alike. The power of the sound in any narrow range of frequencies will be about the same, but the amplitudes and phases of the individual frequency components won't be identical. Nevertheless, the two *sh* sounds will *sound* just the same; we will *hear* them as being the same.

In practice, we use the *ideas* of sine waves and their frequencies and amplitudes to characterize musical sounds, and other sounds as well. The measurements we *really* make are those suitable for our purposes, and are as accurate as they need be. They are similar in concept to Fourier series and Fourier integrals but aren't quite the same. Let us consider how we actually think of and analyze sounds.

We will approach this practical analysis of sounds by means of an idea called *resonance*. Some physical structures respond strongly to a sinusoidal component of a particular frequency. These structures can be used to sort out and measure the frequency components of a sound wave.

In the nineteenth century, the only satisfactory resonators that Hermann von Helmholtz had for his studies of sound were *Helmholtz resonators*. They were usually hollow glass spheres that had two short tubular necks, diametrically opposite one another. One opening was put to the ear, the other directed at the source of a periodic sound. If the sound contained a harmonic whose frequency was equal to or close to the resonant frequency of the cavity of the resonator, the resonator would amplify the harmonic, so that it could be heard separately. The sound in the resonator would also persist after the source of periodic sound had been suddenly turned off. By using a succession of such resonators, Helmholtz could search out and estimate the strengths of the harmonics of a periodic sound. He could also find the frequencies of nonharmonic partials, such as those of bells and gongs.

A symphonic bass is usually bowed.

A jazz bass is usually plucked: jazz bassist Reggie Workman.

We can use a piano to experience something of what Helmholtz did. Hold the forte or sustaining pedal (the "loud pedal") down (this lifts the felt dampers off the strings), and whistle near the strings. After you stop whistling, you will hear a ghostly persistence of the note that you have whistled. This dies away with time. The piano strings act as resonators. Those that can vibrate at the frequencies of the whistled note do so; those that can't, don't.

A taut piano string can vibrate at more than one frequency. The simplest, or *fundamental*, mode of oscillation was described in Chapter 2 (see the illustration on page 28). These vibrations can be regarded as sine waves traveling along the string with a constant velocity $v$, and being reflected repeatedly at the ends of the string. If $L$ is the length of the string, and if the wavelength (distance between peaks of the sine waves) is $2L$, we get the pattern of vibration shown in part A of the diagram on the facing page in which the resonant frequency $f_0$ is

$$f_0 = \frac{v}{2L}.$$

The next three resonant frequencies, $2f_0$, $3f_0$, $4f_0$, have the patterns of vibration shown in parts B, C, and D, respectively.

Plucking or striking a stretched string excites many of these vibrations and produces a complex sound wave made up of many harmonics. The length, mass, and tension of the string determine the periodicity and pitch of the sound that is produced, but do not determine the exact waveform. The relative strengths of the various harmonics depend on whether we pluck or strike a string, and on where along its length we pluck or strike it. We can hear the difference in the sound quality or timbre. For example, the plucked string of a harpsichord sounds quite different from the struck string of a piano. When we excite a string by bowing it, we get a persistent sound of yet another character, even though the resonances of the string may be the same.

Physical resonators are used in some musical instruments. The vertical metal tubes under the wooden bars of xylophones and marimbas, and the bamboo tubes under the brass bars of gamelans, act as resonators that intensify and prolong certain of the partials generated by striking the bars, specifically, the partials that correspond to notes of the scale. Many old instruments, such as the viola d'amore, the viola bastarda, and the baryton, made use of resonators

A

Wavelength = $2L$, frequency = $f_0 = v/2L$

B

Wavelength = $L$, frequency = $2f_0 = v/L$

C

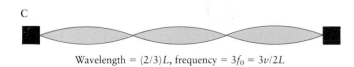

Wavelength = $(2/3)L$, frequency = $3f_0 = 3v/2L$

■ A vibrating string can have several resonant modes, which correspond to standing waves of different wavelengths and frequencies. These are the natural harmonics of the string. Parts **A** through **D** show the wavelength and frequency of the first four harmonics.

D

Wavelength = $(1/2)L$, frequency = $4f_0 = 2v/L$

■ Gamĕlans.

■ The viola d'amore.

■ Jazz trumpeters Howard McGee (right) and Miles Davis: McGee's embouchure is relaxed; he is playing a relatively low note.

■ Jazz trumpeter Woody Shaw: his tense embouchure shows that he is playing a relatively high note.

■ A woodwind: the modern
bassoon.

called *sympathetic strings*. These affected the relative intensities of the partials produced by bowing a string, and emitted a sustained tone even after bowing was stopped or fingering was altered.

The long tubes of brass instruments resonate at harmonically related frequencies. Which one is excited when the instrument is played depends on how the player constricts his lips. The tubular structures of woodwind instruments are resonators whose resonant frequencies are controlled by opening or closing various holes.

Resonances also have a noticeable effect on the timbre of musical sounds. The vocal tract has several resonances that emphasize various ranges of frequency in the sound produced by the vibration of the vocal cords. By changing the shape of the vocal tract, we change the frequencies of these resonances, or *formants,* which determine what vowel sound we produce. The table below gives the first three resonant or formant frequencies for some vowel sounds.

The first three formants of selected vowels.

| Formant number | Vowel | | | | | | | |
|---|---|---|---|---|---|---|---|---|
| | Heed | Hid | Head | Had | Hod | Hawd | Hood | Who'd |
| 1 | 200 | 400 | 600 | 800 | 700 | 400 | 300 | 200 |
| 2 | 2300 | 2100 | 1900 | 1800 | 1200 | 1000 | 800 | 800 |
| 3 | 3200 | 2700 | 2600 | 2400 | 2300 | 2200 | 2100 | 2050 |

The resonances of the soundboard of the violin greatly affect its timbre. The upper graph on the next page shows, as a function of frequency, the ratio of the sound radiated by a good violin to the motion of the bridge caused by bowing the string. The rising and falling curve reduces the intensities of some partials and increases those of others. The suppression of some partials is important for the musical quality of the violin tone. In poor violins, for which a curve like that shown here does not rise and fall so pronouncedly, the sound is harsh.

■ Modern brass instruments.

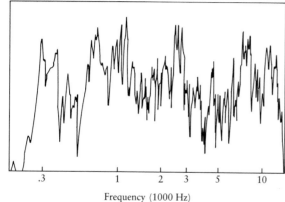

Frequency (1000 Hz)

The body of a violin resonates more at some frequencies than at others. This is a plot of the intensity of the sound wave produced by exciting the bridge of a famous Guarnerius violin with a sinusoidal source of constant amplitude and increasing frequency.

For producing musical sounds, mechanical or acoustical resonators are used. For analysis of sounds, however, we have come to rely on measuring instruments based on electronic filters. When we want to find out what harmonics are present in a sound, we pick the sound up with a microphone and analyze the electrical signal by means of one of the many commercially available electronic filters designed for this purpose. A filter can be described by a curve that shows how the ratio of the output amplitude to the input amplitude varies with the frequency (see the figure at the left below). This filter transmits best at a frequency $f_r$; transmission falls off sharply at lower and higher frequencies.

One electric sound-analyzing device is the *spectrum analyzer*. On a television-like screen, it displays a curve that shows intensity as height, and frequency as distance to the right (as shown in the photograph on the next page). The curve in this figure has several discrete peaks, each corresponding to a particular frequency. Such a graph is called a *line spectrum*; each peak is called a line, and represents a discrete frequency.

In a musical sound, the intensity of the various partials may change with time. This can be depicted by a perspective drawing, as in the right-hand figure on the next page, which shows a sound having six harmonic partials. With the onset of the sound, the intensity of each partial rises to a peak; as the sound dies away, the intensity falls. The higher partials have a lower peak intensity than do the lower partials. As will be seen, such a representation isn't mathematically accurate, because an actual sine wave can't change intensity with time; it goes on at the same intensity forever.

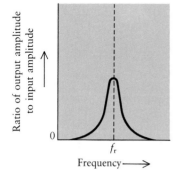

A resonator or electronic filter responds to a narrow range of frequencies only. In this curve the ratio of output amplitude to input amplitude is plotted against the frequency of the input wave. The output peaks the resonant frequency, $f_r$.

The perspective drawing at the bottom of the next page is made up of a sequence of line spectra of a tom-tom. We see that the many frequency peaks or resonances are rather broad.

We can understand why we cannot represent both frequency and time precisely if we consider another sound-analyzing device called a *sonograph* or *sound spectrograph,* which produces sonograms. Frequency is displayed vertically, and time is displayed horizontally; intensity is indicated by the darkness of the shading. The sonograms on page 52 are for a human voice.

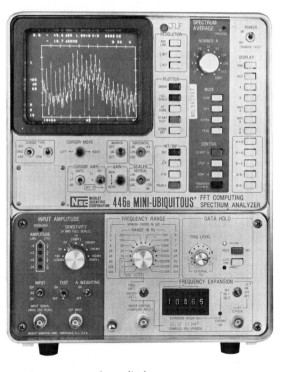

■ A spectrum analyzer displays a graph of intensity against frequency on a cathode-ray tube.

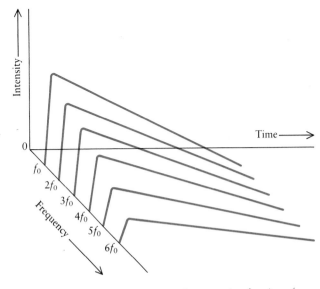

■ A perspective drawing of a sound in which the intensities of the six harmonics shown rise and fall with time.

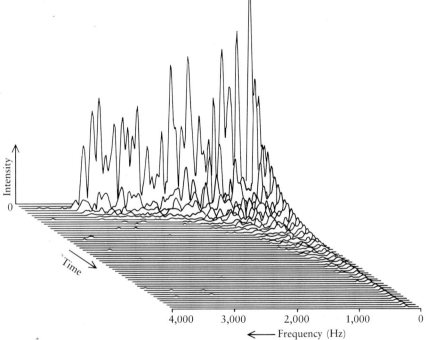

■ Depiction of how the spectrum of the sound of a tom-tom changes with time after the tomtom is struck.

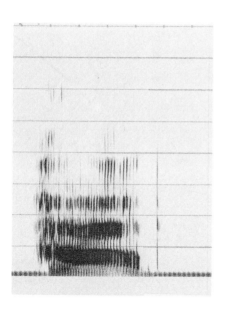

■ Sonograms of the human voice, speaking the vowel *a* in "had."

Both sonograms display broad horizontal bands. These bands represent the resonant frequencies of the vocal tract, or formants, and such resonances change relatively slowly with time as we speak. However, the upper and lower sonograms differ in detail. In the upper sonogram we see horizontal striations. They represent the individual harmonic partials of the voice, and their frequency separation is equal to the pitch frequency. In the lower sonogram we see vertical striations, separated by a time equal to the period, $T$, of the vibration of the vocal cords (the pitch frequency is $1/T$); but we don't see any horizontal striations. The upper sonogram was made with a narrow-band filter, which could sort out individual harmonics. Such a filter cannot respond to rapid changes of sound pressure; so we don't see vertical striations corresponding to individual pitch periods. The lower sonogram was made with a broader-band filter, which responds to several harmonics at once; hence we do not see horizontal striations corresponding to individual harmonics, but we do see vertical striations corresponding to individual pitch periods.

Using filters to analyze sound means that, if we depict frequency in fine detail, we can't depict time in fine detail, or, if we depict time in fine detail, we can't depict frequency in fine detail. This is why the right-hand figure on page 51 is only qualitatively (not quantitatively) correct, for it seems to show both frequency and time very accurately.

Although no musical sound is truly periodic in persisting unchanged forever, most musical sounds are nearly periodic. They can be approximated very closely by a fairly small number of sinelike waves whose amplitudes rise and fall slowly with time, and whose frequencies are nearly harmonically related; so they can be represented by a succession of changing line spectra. The vibration of a piano string dies away slowly; in its line spectrum, the peaks that represent the various harmonics decrease in height as the intensity of the sound decreases. Bells and gongs do not produce periodic sounds, but they do have line spectra that decrease in amplitude gradually as the sound dies away.

Suppose we measure the spectrum of a sound and get a smooth curve rather than a series of spikes, as in the right-hand figure on the facing page. Such a sound can be represented, not by a finite number of frequency components, but by a continuous distribution of partials throughout a range of frequencies. You can make such a sound by whispering a vowel. The frequency (or fre-

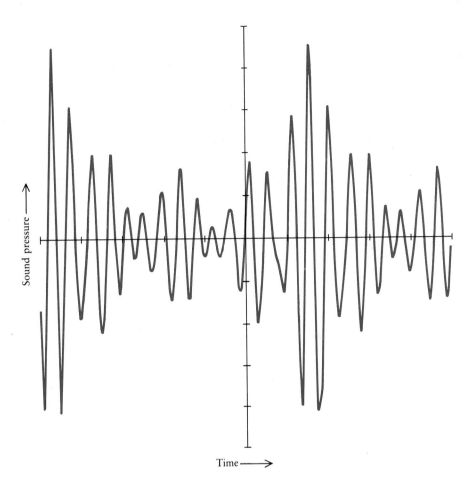

Time ⟶

■ The waveform of a narrow band of noise, a sinelike wave whose amplitude varies randomly with time, but whose frequency is nearly constant.

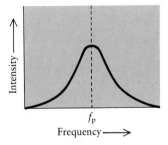

Intensity ⟶

$f_P$

Frequency ⟶

■ Spectrogram of a noise, made up of an infinite number of sinusoidal frequency components whose phases are random and therefore don't peak at the same time. Instead, they give the bell-curve distribution familiar from statistics. A noise with a spectrum like this one will give a sense of pitch, at the frequency $f_P$ at which the spectrogram peaks.

quencies) at which the spectrum peaks depends on the resonance (or resonances) of the vocal tract. If you whisper an *e,* you will get a "high-pitched" sound; if you whisper a *u,* you will get a "low-pitched" sound. This is because the resonances of the vocal tract occur at higher frequencies for an *e* than for a *u.* In a whispered *e,* the resonances emphasize high-frequency components in the breathy sound. In a whispered *u,* the resonances emphasize lower-frequency components.

Noise can be narrow-band or broad-band; that is, the peak, or hump, in the spectrum can be narrow or broad. If there is no peak in the spectrum, so that all frequencies within a range are present equally, the noise is called *white noise.* This is the even, breathy, faintly frying sound that you get on some noisy telephone connections, or when you turn the volume control of an AM radio up when it is not tuned to a station.

If we gradually narrow the width of a spectral peak like that shown at the right above, we will get a more and more pronounced sense of pitch, which corresponds to the frequency ($f_P$) of the peak. When the peak becomes very narrow, the noise ceases to sound noisy. Instead, it sounds like a sine wave that wavers in amplitude, and a little in frequency (see the left-hand figure on the preceding page). Max Mathews (in *The Second Law*) and others have produced computer music that uses noise of various bandwidths, both narrow-band tonelike noise and broader-band shushing noise.

Sounds that have a continuous spectrum are not necessarily noises. If the phases of all the sinusoidal frequency components are equal, so that they all peak at the same instant, or if the phases change slowly and smoothly with frequency, we get a single pulse rather than a chaotic, persistent noise.

Part A of the upper figure on the facing page shows a short pulse or burst of pressure that would sound like a click, and part B shows that the sinusoidal components of which it is composed span a broad range of frequencies, from 0 on up. Part C shows a pulse that has a broader rise and fall of amplitude. It will be more thumplike than clicklike, and part D shows that its spectrum spans a narrower range of frequencies. In general, the shorter the pulse of sound, the broader the range of frequencies. Roughly, the width of the frequency range or band ($B$) is inversely proportional to the duration ($D$) of the sound in seconds:

$$B = 1/D.$$

Most short sounds aren't simple rises and falls in pressure like those in this figure. When we strike a block of wood, we get a sort of click, but the pressure tends to oscillate up and down for a short time. Similarly, when we strike a bass drum, we get a sort of thump, but the drum head oscillates for a short time. The bandwidth or range of frequencies of such sounds is still proportional to the duration of the sound, but the peak of the spectrum is at the frequency with which the struck object oscillates: the larger the object, the lower this frequency of oscillation, and the duller the sound. For example, in part A of the lower figure on the facing page, we have a short section of a waveform, two cycles long, and in part B a longer section, four cycles long, twice the duration. Such short sections of sine waves will be heard as short beeps rather than as clicks. In both parts of the figure, the peak of the spectrum is at the frequency $f_0$ of the sine wave, but the spectrum of the longer, four-cycle section is half as wide as the spectrum of the short, two-cycle section.

When a sine wave is turned off or on abruptly, this excites resonant devices of many frequencies. Indeed, we hear a click when the wave is turned off or on

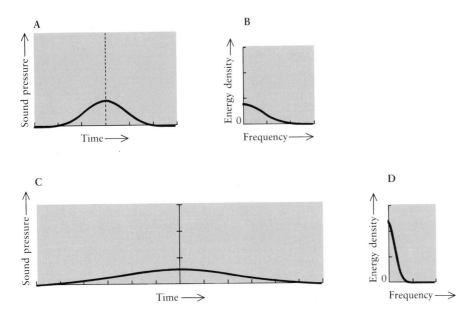

■ The spectrum of a single pulse of air pressure is not a line spectrum, but a continuous spectrum made up of components of a continuous range of frequencies. If the pulse is short (part **A**), the spectrum is broad (part **B**), and the pulse sounds like a click. If the pulse is longer (part **C**), the spectrum is narrower (part **D**), and the pulse sounds like a thump.

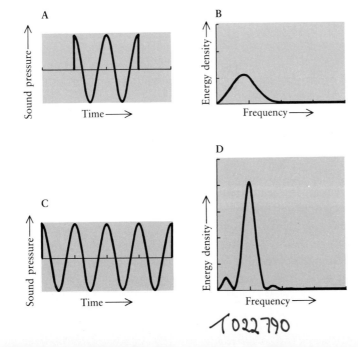

■ A short section of a sine wave of frequency $f_0$, as in part **A**, has a broad spectrum centered on the frequency $f_0$, as in part **B**. A longer section of the same sine wave, as in part **C**, has a narrower spectrum, as in part **D**, still centered on the frequency $f_0$. When the sine wave is turned on abruptly, as shown here, we hear a click.

abruptly. Anyone who listens to the output of an audio oscillator, and breaks the connection with the speaker by flipping a switch, will hear such a click. We can get the effect of hearing a short pure tone or sine wave without a click by increasing and decreasing the intensity gradually, as shown in the figure on the facing page. Of course, this "sine curve of changing intensity" isn't truly a sine wave, because a sine wave persists forever with the same intensity. But a variation of sound pressure, such as that shown in the figure, *sounds* like a pure tone, except that the loudness rises and falls, and measuring instruments respond to it much as they would to a sine wave, except that the response goes up and down as the intensity rises and falls.

The practical result here is that we can add "slowly changing sine waves" to produce periodic sounds of finite duration. Mathematically, such a "slowly changing sine wave" is not a sine wave at all, but it approximates many of the properties of a sine wave. Measuring instruments and the human ear respond to such a "slowly changing sine wave" as if it were a pure tone that changes slowly in amplitude, frequency, or phase. However, if we change sine waves too fast, we will hear a click or the twang of a plucked string.

On the screen of a spectrum analyzer, a pure tone of slowly increasing amplitude appears as a line or peak of increasing height, and the corresponding sound grows louder and louder without changing pitch (much). If we change the frequency of a sine wave gradually, the line on the spectrum-analyzer display moves to the left or right, and we hear a decrease or increase in pitch. A periodic change in frequency (a vibrato) is heard as a periodic change in pitch if it is slow, and the line on the spectrum-analyzer display jiggles back and forth. Above a vibrato rate of about six per second, we can no longer hear vibrato as a change in frequency; rather, vibrato gives an entirely new and pleasant quality to a musical sound.

All that I have said about a single slowly changing sine wave also applies to sums of slowly changing sine waves. Combinations of sine waves, each of which changes slowly in amplitude or frequency, are heard as tones of changing loudness or pitch. In the chapters that follow, many sounds will be described as combinations of sine waves or partials, even though the amplitudes and frequencies of the partials change with time.

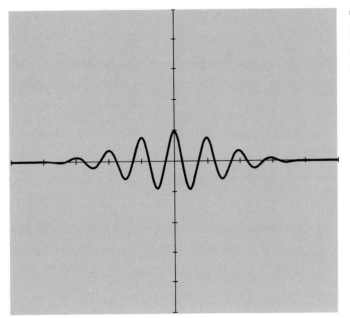

■ If a sine wave is turned on and off slowly, we don't hear a click. (Mathematically, this is not a sine wave but, for practical purposes, it sounds like one.)

# A Note on **Sine Waves**

A sine wave is not just any wiggly curve that rises and falls with time; it is a precise mathematical function that can be described very simply. The crank illustrated in part A of the figure at the left turns with a constant speed, making one revolution every $T$ seconds. The height of the crank at any particular time is shown as $h$. We can draw a sine wave by plotting the height of the crank against time.

To begin to do this, as shown in part B of the figure, place equally spaced points around the circumference of a circle. The figure shows eight points, numbered from 0 to 7. For each successive point, starting with 0, measure the height of the point above or below the horizontal axis line through the center of the circle. Plot the successive heights of points 0, 1, 2, 3, and so forth, as in part C, at successive equally spaced distances along the horizontal line; then connect these points with a smooth curve to draw the sine wave. If you use more equally spaced points around the circle, you can draw the sine wave more accurately, as in part D.

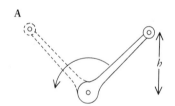

**A**

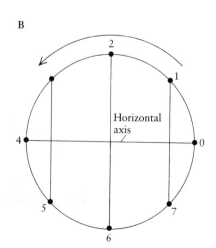

**B**

Horizontal axis

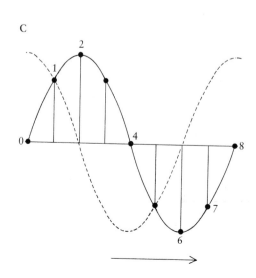

**C**

■ **A.** The height, $h$, of the handle of a crank that rotates at a constant speed varies sinusoidally with time.
**B.** Equally spaced points around the circumference of a circle.
**C.** When the points in part B are translated to a graph in which the steady passage of time is represented as movement to the right, they outline a sine wave.
**D.** A smooth sine wave would be produced if every point on the circle were translated to the graph.

Mathematically, the relation between $h$ (or amplitude) and time $t$ is written $h = \sin wt$, in which $w$ is a constant, equal to $2\pi$ times the frequency $f$, and $f = 1/T$, $T$ being the period. The higher the frequency (the shorter the period), the more times the sine wave goes up and down each second.

There is another sine wave, the *cosine,* which is just a sine wave with a different phase. The cosine of $wt$ is written $h = \cos wt$. The cosine reaches its peak a quarter of a period ($T/4$) before the sine does. The cosine is traced out by the height of the dashed crank in part A of the figure, which is at right angles with the solid crank, and is shown as the dashed curve in part C. To draw the cosine we simply start at point 2 instead of point 0, and plot successively the heights of points 2, 3, 4, 5, 6, and so forth.

This simple "mechanical" picture can tell us all we need to know about sine waves. Those who like mathematical puzzles can work out all sorts of interesting, useful, and important relations, such as

$$(\sin wt)^2 + (\cos wt)^2 = 1.$$

A particularly interesting and important relation is

$$(\sin pt)(\sin wt) = (\tfrac{1}{2})[\cos(w + p)t - \cos(w + p)t].$$

Here we can think of the sine wave $\sin pt$ as controlling the intensity or amplitude of the sine wave $\sin wt$. What can we discover from this simple relation? First, we see that the product of two sine waves is simply two sine waves (since cosine waves are just sine waves that peak at a different time, that is, have a different *phase*). The average value of the product of two sine waves is zero, because the average value of each of the two sine waves is zero. If the sine waves have exactly the same frequency, one of the two components on the right is the cosine of zero, which is equal to one. In mathematically representing a function of time as a sum of sine waves, we make use of this property.

We can deduce another important result thing from this last formula. Imagine that $p$ is very small compared with $w$, say, $p = 1$ and $w = 100$. Then the factor $\sin pt$ will cause the amplitude of the oscillation $\sin wt$ to vary slowly with time. Our formula tells us that

$$(\sin t)(\sin 100t) = (\tfrac{1}{2})(\cos 99t) - (\tfrac{1}{2})(\cos 101t).$$

That is, a slowly varying sine wave doesn't have just one frequency; mathematically it is made up of two or more frequencies, but these frequency components are very near the nominal frequency of the slowly varying sine wave. This is very important, for when we turn a sine wave on or off, we still think of it as having the same frequency that it would if it persisted forever. In practice, to our sense of hearing, this is true *as long as we turn the sine wave on and off slowly enough.*

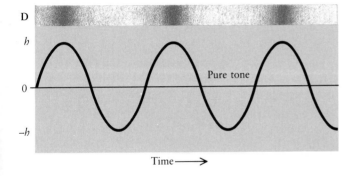

D

$h$

0

$-h$

Pure tone

Time ⟶

# Scales and Beats

At one time or another many of us have heard a piano tuner tuning a piano. We have heard him strike an octave and then use his wrench (or key) to tighten or loosen a string. How does he know when the octave is tuned just right? The *beat* between the two notes disappears. What is this beat that guides the tuner in his tuning?

The figure on the facing page shows the sum of two sine waves of slightly different frequencies. This sum looks like a single sine wave whose amplitude increases and decreases slowly with time, and that is what it sounds like. The two sinusoidal components seem to merge into a single throbbing or beating sound. If we make the sine waves more and more nearly equal in frequency, the beat between them becomes slower and slower. When the sine waves have equal frequencies, the beat disappears, and we hear a tone of constant amplitude.

What beat does the piano tuner listen for in tuning an octave? A single note of a piano has many harmonics, and its second harmonic is an octave higher than its fundamental tone. In tuning an octave, the piano tuner listens for beats between this second harmonic of the lower note and the fundamental of the higher note. When this beat disappears, the octave is in tune.

Actually, piano strings have a little stiffness, which adds to the effect of tension in keeping the strings straight. As a result, the higher partials have frequencies that are a little greater than integer multiples of the fundamental frequency $f_0$; that is, they are slightly larger than $2f_0$, $3f_0$, $4f_0$, $5f_0$, and so forth.

When octaves in a piano are tuned by the method of eliminating beats, the octaves will therefore be *stretched* a very little; that is, they will have frequency ratios slightly greater than 2. Pianos *are* often tuned with slightly stretched octaves, sometimes because of the stiffness of the strings, sometimes because the pianist prefers the brighter timbre that results. In this discussion, we will ignore the stiffness of piano strings and other practical realities, and assume that *all* musical tones have partials that are exactly harmonic; that is, that are integer multiples of the frequency of the first partial.

Intervals other than octaves also can be tuned by means of beats. To see how, consider the piano keyboard, shown on page 64. We go up a *semitone* when we go from any key (white or black) to the next key (black or white). Certain intervals are called *consonant intervals,* because musicians find that notes sepa-

**A**

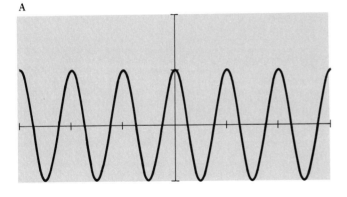

**B**

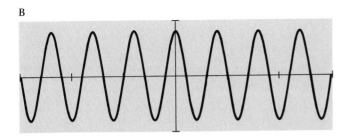

**C**

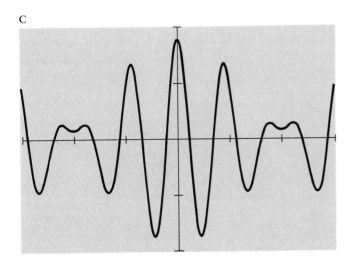

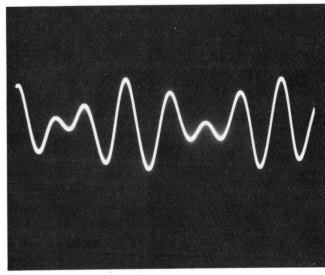

■ The phenomenon of beats. The amplitudes of the two sine waves in parts **A** and **B**, whose frequencies are very close to one another, add up to give the wave in part **C**, which looks very much like a sine wave of slowly varying amplitude. The photograph of an oscilloscope screen in part **D** shows the beat between tones of 4,000 and 4,100 Hz. When the frequency difference is small, it sounds like a sine wave or pure tone whose intensity rises and falls slowly with time. As the frequencies of the sine waves are made more and more nearly equal, the rate at which the amplitude rises and falls (the frequency of the beat, or the beat frequency) gets slower and slower until, when the frequencies are equal, the beat disappears.

The white keys of the piano give the seven notes of the C-major diatonic scale.

rated by these intervals sound pleasing when sounded together. These intervals are given in the table below, which also gives the ideal frequency ratio between the two notes and the number of semitones of difference between them.

The figure at the top of the next page shows the harmonics of a C with a frequency $f_0$ and the G above it, which has a frequency $(3/2)f_0$. We see that the third harmonic of C has the same frequency, $3f_0$, as the second harmonic of G. If the C and G are a little mistuned, these harmonics will produce an audible beat when the two notes are struck together. By tuning G so that it doesn't beat with C, we can assure that the fundamental frequency of G is just 3/2 that of C, a perfect fifth.

Consonant intervals.

| Name of interval | Notes (in key of C major) | Ideal frequency ratio | Number of semitones |
|---|---|---|---|
| Octave | C–C | 2 | 12 |
| Fifth | C–G | 3/2 | 7 |
| Fourth | C–F | 4/3 | 5 |
| Major third | C–E | 5/4 | 4 |
| Minor third | E–G | 6/4 | 3 |
| Major sixth | C–A | 5/3 | 9 |
| Minor sixth | E–C | 8/5 | 8 |

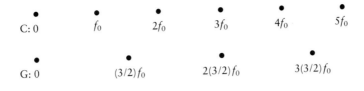

■ The frequencies of two notes (such as C and G) that are an interval of a fifth apart are always in the ratio of 100:150. Their harmonics will be in the same ratio; so the second, fourth, and so forth, harmonics of G will coincide with the third, sixth, and so forth, harmonics of G, because at each of these coincidences $3f_0 = 2(3/2)f_0$.

A little thought will show that other notes can be tuned in this way as well. A fourth has a frequency ratio of 4/3; so the third harmonic of F should have the same frequency as the fourth harmonic of C (which is C″, two octaves above the first C).* A major third has a frequency ratio of 5/4; so the fourth harmonic of E should have the same frequency as the fifth harmonic of C. A minor third has a frequency ratio 6/5; so the fifth harmonic of E should have the same frequency as the sixth harmonic of C (a G).

A friend of mine once tried to tune his piano according to such relations, thinking that by later adjustments he could get to the *equal-tempered* scale, to which a piano is really tuned. He underestimated the difficulty. Piano tuners have a systematic method of tuning, in which they tune the intervals within a single octave to have a prescribed number of beats per second and, once they have tuned these twelve notes, they tune all the other notes in octaves above and below by the method of octaves, with no beats. This isn't what my friend did.

Here we come to the dilemma of the diatonic scale, the notes of the white keys of the piano. Is there a *sensible* explanation for this scale, which is used in so many cultures? There *is* an explanation, but you must judge for yourself how sensible it is.

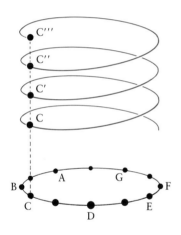

■ The notes of the diatonic scale as points along an ascending spiral. C above middle C is in some ways very like middle C, in some ways a little different. Here all Cs are close together and hence alike; all Ds are close together and hence alike; and so on.

First, we should observe that notes or tones an octave apart sound very similar. In primitive music, men and women sing in octaves without realizing that they are singing different notes. For some timbres, it is not uncommon to make an error of an octave in judging pitch. The psychologist Roger Shepard has likened the similarity of octaves to points on an ascending spiral. As shown in the adjacent figure, the notes C, D, E, F, G, A, B, C repeat again and again around the spiral as we go up the scale. But, C′ an octave up always lies close (in perception) to C an octave below, and so it is for all the other notes of the scale.

---

* The primes are used to flag *relative* positions of notes. As we start up, we have the notes C, D, E, F, G, A, B; in the next octave up the notes are C′, D′, E′, F′, G′, A′, B′; in the next, C″, D″, and so forth. This convention helps us keep track of *which* C we are talking about as we jump up and down the octaves in a discussion of harmonics, say. The notation is *relative*, not absolute; any note we choose can be the starting point for the primes in a given discussion.

With this in mind, consider the figure below, which shows notes on the bass and treble musical staffs. In part I, we see the first six harmonics of the C an octave below middle C.* They are, successively, (middle) C′, G′, C″, E″, and G″. The intervals between these successive notes are consonant intervals. They are the octave (C to C′), the fifth (C′ to G′), the fourth (G′ to C″), the major third (C″ to E″), and the minor third (E″ to G″).

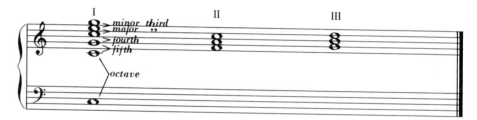

■ The successive harmonics of a note C give us the notes C′, G′, C″, E″, and G″, and include the intervals of the octave, the fourth, the major third, and the minor third. In any octave, if we go up from G by a major third, we get B, and if we go up from B by a minor third, we get D. If we go down from C by a fifth, we get F, and if we go up from F by a major third and then by a minor third we get, successively, A and C. We now have all the notes of the diatonic scale.

So far we have only three of the seven notes of the diatonic scale, C (the *tonic*), E (the *mediant*), and G (the *dominant*). A plausible way in which to get one more note is to drop down a fifth from C″, as shown in part II of the figure. This gives us F′ (the *subdominant*); the frequency ratio of F′ to C′ is 4/3.

In going from C″ to E″ to G″, as in part I, we go up first by a major third (a frequency ratio of 5/4), then by a minor third (a frequency ratio of 6/5), and so arrive at an overall frequency ratio between G and C of 3/2.

The fifth or dominant, G, plays an important role is music. It seems reasonable to go up from G in the same pattern, first by a major third, then by a minor third, as in part III, thus basing a *major triad* on G. This gives us the seventh, or *leading*, tone, B, and the second, D. The fourth, F, the subdominant, also plays an important part in music. Let us also base a major triad on F, again going up by a major third, then by a minor third, as shown in part II. This gives A (the *submediant*) and brings us back to C. Here A′ is a minor third below C″; that is, (6/5) A′ = C″ and, because C″ = 2 C′, the frequency ratio of A′ to C′ is 5/3.

We now have all seven notes of the diatonic scale in the key of C; C, being the first note of the scale, is the *tonic*.

* It is called "middle C" because it is written on the line in the middle, between the treble and bass clefs.

Musical intervals with frequency ratios given by the ratio of integers, such as 3/2 (fifth) and 5/4 (major third), are the basis of the diatonic scale. Such musical intervals are, of course, important in music itself. Let us concentrate on the black notes in the short passage of music shown here. They are C, A, D′, G, C. Let us call the frequency of C, $f_0$. In going to A, we go to a frequency of $(5/3) f_0$. The interval from A to D′ is a fourth, with a frequency ratio of 4/3; so when we arrive at D′, the frequency should be 4/3 that of A, which is 5/3 times $f_0$, or $(4/3) (5/3) f_0 = (20/9) f_0$. We now go down by a fifth to G, which should give us a frequency of $(2/3) (20/9) f_0 = (40/27) f_0$. We now go down another fifth to C, which should give us $(2/3) (40/27) f_0 = (80/81) f_0$. But we started out by setting the frequency of C equal to $f_0$, and we haven't gotten back there!

■ Consider the black notes in this series of chords. They proceed from C to A to D′ to G to C, or by intervals of a sixth up, a fourth up, a fifth down, and a fifth down. If these were perfect intervals, in which the frequency ratio of the sixth is 5/3, of the fourth is 4/3, and of this fifth is 3/2, we would not get back to the same pitch from which we started.

What this short passage of music shows is that the diatonic scale does not proceed from one note to another by means of intervals whose frequency ratios are ratios of whole numbers. In progressing up and down by means of such ideal intervals, we would always be forced to wander away from any fixed scale.

In a very real sense, perfect intervals whose frequency ratios are the ratios of integers are the very foundation of music. These intervals derive from the harmonics of a musical tone. They are important to the ear. So is the diatonic scale, which we have explained in terms of these perfect musical intervals; yet the perfect intervals and the diatonic scale are inherently at odds. What do we do?

One way out is called *equal temperament,* which we use in tuning the piano. The concept of temperament can be conveniently explained in terms of a musical interval called a *cent.* The cent is a hundredth of a semitone, or $1/1{,}200$ of an octave. The frequency ratio of the cent is the $1/1{,}200$ power (the 1,200 root) of 2, or 1.00057779. If we multiply 1,200 of this decimal number together (that is, if we raise it to the 1,200 power), we get 2, the frequency ratio of the octave. If we raise the number to the 100 power, we get the frequency ratio of the equal-temperament semitone, 1.059463.

The intervals of differently tempered scales, measured in cents.

| Note | Equal-tempered | Pythagorean | Just |
|------|----------------|-------------|------|
| C | 0 | 0 | 0 |
| D | 200 | 204 | 204 |
| E | 400 | 408 | 386 |
| F | 500 | 498 | 498 |
| G | 700 | 702 | 702 |
| A | 900 | 906 | 884 |
| B | 1100 | 1100 | 1088 |
| C | 1200 | 1200 | 1200 |

The table above shows the intervals of the diatonic scale in cents, for equal-tempered tuning, for *Pythagorean* tuning (which is designed to give perfect fifths), and for *just* tuning. The table below shows the "error" in these compromise tunings relative to the ideal ratios.

Some musicians love just temperament dearly. Harry Partch, who made many strange instruments whose sounds were as wonderful as their names, and which were often of somewhat uncertain pitch, had a harmonium justly tuned in the key of C. It sounded excellent in C, but dreadful when played in any other key.

Errors in tempered scales relative to ideal ratios.

| Interval | Name | Equal-tempered error | Pythagorean error | Just error |
|----------|------|----------------------|-------------------|------------|
| C–E | Major third | 14 | 22 | 0 |
| D–F | Minor third | 16 | 22 | 22 |
| C–F | Fourth | 2 | 0 | 0 |
| C–G | Fifth | 2 | 0 | 0 |
| C–A | Sixth | 16 | 22 | 0 |
| D–A | Fifth | 2 | 22 | 22 |

Let us consider just temperament closely. The musical example on page 67 shows that in just temperament the musical interval of a fifth from D to A is in error by 22 cents; that is why we are able to get back to C in just temperament, but at a considerable cost. This 22-cent error is 6 cents larger than any error in equal temperament.

Furthermore, in equal temperament we have dealt very simply with the problem of sharps and flats. Ideally, in the interval of the diminished sixth, A ♭ should have a frequency ratio of 1.6 to be a perfect major third from the C above it. In order to make the interval E–G ♯ a perfect major third, G ♯ should have a frequency ratio 25/16 = 1.5625. In our equal-tempered scale, both A ♭ and G ♯ are represented by the same black key, and by a frequency ratio of 1.5874.

There are flats and sharps in diatonic scales in which the tonic or first note is other than C. A fifth above C is G, the first note of a closely related key, G major, which has six notes in common with C major (in the key of G major we use F ♯ instead of F). A fifth above G is D, and the scale starting on D has two sharps, C ♯ and F ♯, and has only five notes in common with the C scale.

Going up through a cycle of successive fifths on the piano, we reach all keys, white and black, and on each we can build a scale. Or we can start down from C by an interval of a fifth, and reach F. The scale in which F is the first note, or tonic, has six notes in common with the C scale; in F, B ♭ is used instead of B.

Bach was one of the first strong advocates of equal-tempered tuning. In equal temperament, the intervals, though slightly "in error," are the same in all keys. Bach's *Well-Tempered Clavier* includes pieces in all keys. With equal-temperament tuning, all are equally in tune (or out of tune).

It is often argued that equal temperament can be offensive to musicians with keen ears. Helmholtz inspired the production of keyboard instruments with almost unplayably complex keyboards in order to allow the production of close approximations to perfect (integer frequency ratio) intervals. In fact, most people, including some musicians, find it difficult to tell perfect intervals from equal-tempered intervals. Max Mathews prepared a computer-generated tape that included both equal-tempered and perfect chords, and equal-tempered and perfect intervals. A good many people couldn't distinguish them, but one musician could infallibly tell an equal-tempered fifth from a perfect fifth, a difference of only 2 cents.

There has been much discussion of whether or not string players play perfect or tempered intervals when playing solo. Such measurements as have been made indicate that they play neither. And the musical passage on page 67 proves that they couldn't play a succession of perfect intervals and get back to the pitch of the note on which they started.

So far we have considered the frequency ratios between the notes of a scale. There is another characteristic of a scale. On which note does it start?

The diatonic C-major scale, represented by the white keys of the piano, consists of seven notes, or eight if we span a whole octave. Within the octave of the major scale, five of the intervals are whole tones (two semitones), and two (the third and the seventh) are semitones. The succession of intervals that we will encounter in going up the scale depends on which note we choose as the beginning of the scale, and sharps or flats are added to preserve the sequence of whole tones and semitones, since this sequence defines the scale.

In classical Greek music, the scale could start on any note of the diatonic scale (the white keys of the piano), and the semitones fell wherever they fell; such scales were called *modes*. The Greek modes vanished from music during the early centuries of the Christian era, and were replaced by the Church modes, which flourished from 800 to 1500. These modes were the basis of Gregorian chant. They carried over into contrapuntal compositions such as those of Palestrina. Although they fell largely into disuse in the seventeenth century, something much like these modes persists in the folk music of various countries, and some serious composers still write modal works or passages. But generally in Western music, only two modes survive, those that proved best suited to Western harmony: the *major* scale and the *minor* scale. If we start on C on the piano and play the white keys to C′, we play a major scale, in which the pattern of whole tones (2) and semitones (1) is 2, 2, 1, 2, 2, 2, 1.

If we start on A, and play the white keys to A′, we play a minor scale, in which the interval pattern is 2, 1, 2, 2, 1, 2, 2. There are three variants on the minor scale as well, in which the last three intervals in the pattern can be 2, 1, 2, or 2, 2, 1, or even 1, 3, 1, produced by using (in the key of A minor) F♯, or G♯, or both, instead of F and G.

In this book, the central concern is with the science of musical sound. In this chapter, it has been noted that the musical intervals produced by instruments can be tuned by listening for beats between the harmonics of two notes that are sounded simultaneously. We have seen that one cannot choose the frequencies of a seven-tone (or twelve-tone) scale in such a way that the ratios of the frequencies of the notes of the intervals are ratios of integers. We have seen that equal-tempered tuning, based on semitones of frequency ratio 1.059463, gives excellent approximations to ideal intervals, and that these approximations are the same for scales in all keys.

When we venture beyond the matters of intervals and tuning to discuss modes, we are wandering a little beyond the science of musical sound, into the territory of music itself. This we cannot wholly avoid. But we will concern ourselves only with the two surviving modes: the major scale, because it is important to Rameau's ideas concerning harmony; and the minor scale, because it is hard to avoid.

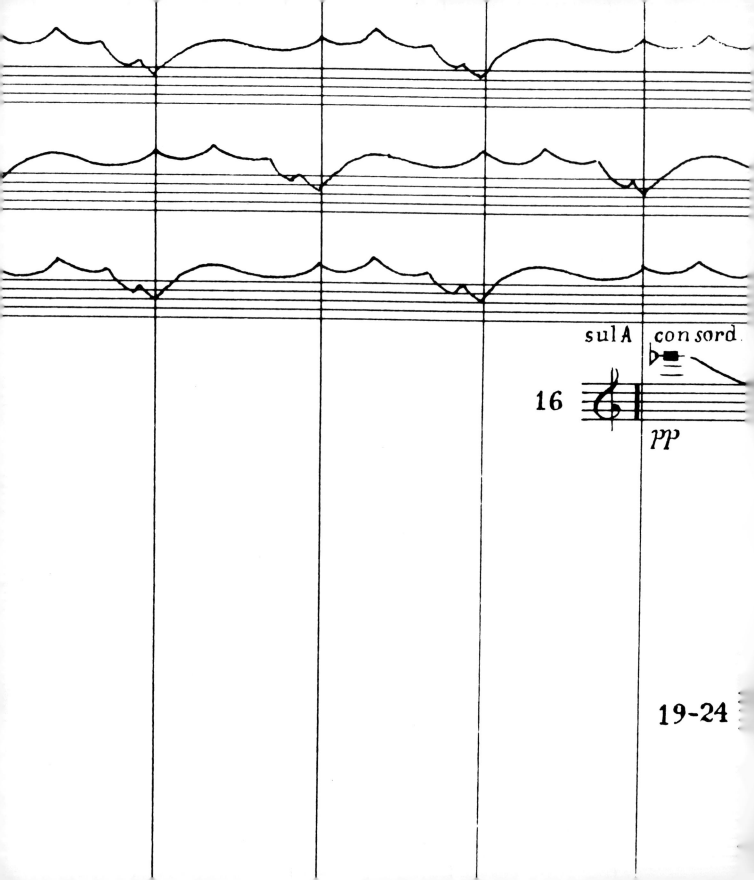

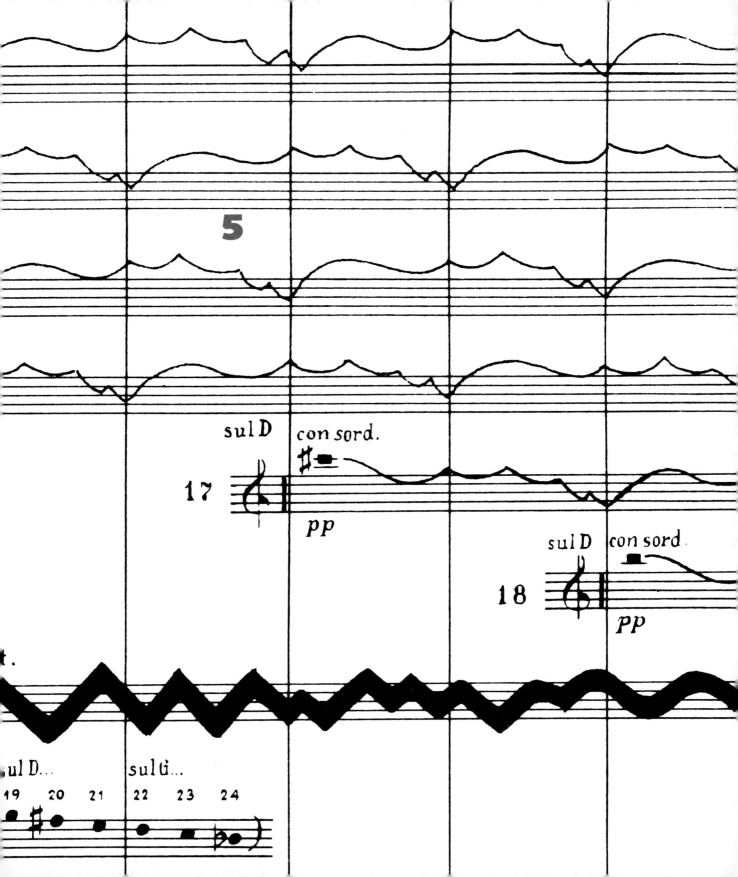

# Helmholtz and Consonance

Melody is characteristic of almost all music. Strangely, so is the diatonic scale (or some subset of it), with intervals based on frequencies in the integer ratios 2/1, 3/2, 4/3, 5/4, 6/5, and so on. No doubt, musicians of many ages and cultures have plucked strings simultaneously and tuned their instruments by observing beats, as piano tuners do today. However, sounding notes together in music in what we conceive of as harmony is peculiar to Western music.

The notes of one instrument or several can be sounded together for a variety of purposes. An orchestra can make a frightful din. In giving instructions for the representation of cannon shots in his piece *Les Caractères de la guerre*, written in 1724, François Dandrieu recommended that the harpsichordist "strike the lowest notes on the keyboard with the entire length of the hand." By 1800, another French composer recommended striking the lowest three octaves with the flats of both hands in his rendition of the sound of a cannon.

Whatever we think of as the essence of harmony, it does not lie in such sounds, or in the near-tone clusters of Percy Grainger, the tone clusters of Henry Cowell, or even the haunting reminiscence of church bells that Charles Ives evokes in his *Concord Sonata* by simultaneously striking many black keys softly with a bar of wood.

We think of harmony as notes that are sounded together smoothly and sweetly, or as rough-sounding combinations of notes, full of tension, that miraculously resolve into a succeeding consonant chord. Beyond this, we think of well-known (and sometimes well-worn) progressions from chord to chord that serve as phrases or words of music. We think of modulation, the shifting from key to key, which is sometimes forthright, sometimes elusive or ambiguous.

Of the harmonic language of music this book has little to say. Our concern will be chiefly with the consonance or dissonance one experiences when two, three, or more notes are sounded together. It is these experiences of consonance and dissonance that underlie the evolution of the musical theory of harmony.

Of consonance, dissonance, and chords, there can be several views. One might be that we simply learn to regard certain combinations of notes as consonant, others as dissonant. Composers of our century present us with an extraordinary range of combinations of notes, and with some rules for making such combinations. Can consonance and dissonance spring from nothing more than rules and customs? As will be seen, in a sense they can.

**5**

■ The score on the preceding pages is from *Polymorphia,* for forty-eight stringed instruments, by Krzysztof Penderecki.

■ Page from score of Ives's *Concord Sonata*, showing instructions for playing black keys with a stick.

I believe, however, that to look at experiences of consonance and dissonance as arising from rules is to look at things the wrong way round. Rather, I believe that the rules and customs are based on experiences of consonance and dissonance that are inherent in normal hearing. Of course, musical training and sophistication will change the subjective experience. The trained ear hears much in harmony that escapes the musically untrained. Sometimes the trained ear hears things that aren't there, for a musical friend of mine hears G♯ and A♭ as different on the piano. All this we will consider later.

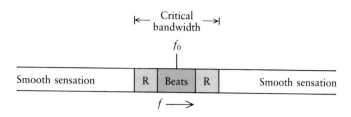

■ Pure tones (sine waves) that are close enough together in frequency give rise to audible beats. Even when the frequencies of such tones are too far apart for us to hear beats, we can still hear a rough sensation. If the frequencies differ still more, we hear each tone separately and smoothly. Imagine one pure tone to have a frequency $f_0$ and another to have an adjustable frequency $f$. As we vary $f$ from much below $f_0$ to much above $f_0$ we pass from smooth to rough to beats to rough to smooth again, as shown. The range of frequencies within which we hear roughness or beats is called the *critical bandwidth*.

Let us return to the piano tuner. Beats and "roughness" are phenomena that are crucial to consonance and harmony. We are already acquainted with beats. We know that the piano tuner tunes intervals for the absence of beats, or else for the number of beats per second that leads to the equal-tempered scale. In the nineteenth century, Helmholtz tried to explain consonance and harmony entirely in terms of beats. He thought that intervals were consonant if there were no (or few) beats between their partials. For dissonant intervals, he proposed that partials of different tones were so close together in frequency that the beating between them was perceived as dissonance.

The work of Plomp (1976) and others has shown that this is too simple a view. Slow beats do not give a sense of dissonance, but merely a tremolo, a rising and falling of amplitude. Further, as we gradually separate the frequencies of two sine waves or "pure tones," we hear a disagreeable roughness even when the frequencies are so far apart that we no longer hear beats. As shown in the figure above, the range of frequencies in which we hear beats or roughness is called the *critical bandwidth*.

The critical bandwidth is an important experimental fact of hearing. To some degree, in listening to sounds we can tune in on a narrow band of frequencies, much as we tune in to a radio channel. When frequency components are separated by more than a critical bandwidth, we can hear them separately (this is called *hearing out*). But frequency components that lie within a critical bandwidth interact, and give us sensations of beats, roughness, or noise.

The critical bandwidth is important in the perception of loudness, in the perception that a sound is noise, and in the *masking* or hiding of one sound by another. In essence, the critical bandwidth results from the way that the ear resolves frequencies. At low and moderate sound levels, frequency components lying farther apart than a critical bandwidth send signals to the brain over separate nerve fibers, but frequency components lying within a critical bandwidth send a mixed signal over the same fibers.

The graph on the facing page shows the relation between critical bandwidth and consonance in another way. We can see from it that the maximum dissonance occurs at about a quarter of a critical bandwidth. For greater frequency separations, consonance increases, and becomes almost perfect for *all* separations greater than a critical bandwidth.

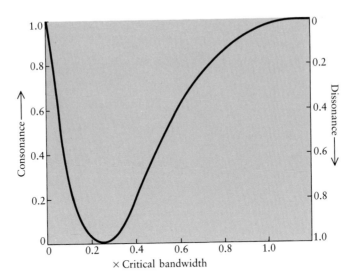

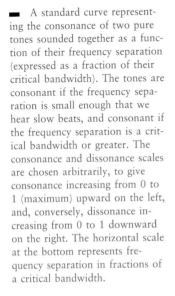

A standard curve representing the consonance of two pure tones sounded together as a function of their frequency separation (expressed as a fraction of their critical bandwidth). The tones are consonant if the frequency separation is small enough that we hear slow beats, and consonant if the frequency separation is a critical bandwidth or greater. The consonance and dissonance scales are chosen arbitrarily, to give consonance increasing from 0 to 1 (maximum) upward on the left, and, conversely, dissonance increasing from 0 to 1 downward on the right. The horizontal scale at the bottom represents frequency separation in fractions of a critical bandwidth.

In order to know what frequency intervals between pure tones are consonant, we must know how the critical bandwidth varies with frequency. This is shown in the graph on the next page. Vertically we show, for a critical bandwidth, the difference between the two frequencies that lie at its edges, $f_1$ and $f_2$. Horizontally we show the average value of two frequencies, that is, $(f_1 + f_2)/2$.

We see that, for most of the frequency range shown, the critical band lies between a minor third and a whole tone. For frequencies below 440 Hz (the A above middle C), the critical bandwidth is larger. Thus we might expect that, in order to be consonant, notes struck together at low frequencies would have to be more widely separated than notes struck together at high frequencies. Indeed, in piano music the notes of the chords in the bass are commonly more widely separated than the notes of the chords in the treble. However, for many musical sounds lower frequencies cannot be very important to consonance, since we hear mostly the higher partials, partly because a good deal of the energy is in the higher partials (almost all when we hear via a transistor radio), and partly because the ear is more sensitive to higher frequencies than to lower frequencies.

A good rule of thumb is that pure tones less than a minor third apart are dissonant, but tones a minor third or more apart are consonant; so, for pure sine tones, *any* interval greater than a minor third will be judged as consonant, however odd the ratio of frequencies. But, of course, this isn't so for musical tones, such as piano tones, which have many harmonic partials. And musicians tend to judge the consonance or dissonance of pairs of sinusoidal tones by first recognizing the musical interval, and then calling the pair of tones consonant or dissonant on the basis of their past experience with pairs of nonsinusoidal tones, or on the basis of what they have been taught.

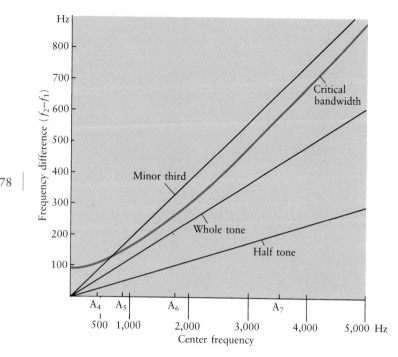

The width of a critical band-width as a function of frequency. Above a frequency of 500 Hz the critical bandwidth is roughly proportional to frequency. The frequency differences for three intervals are shown. A good approximation is that, if two pure tones are separated by a minor third or more, they will sound consonant together.

The graph at the top of the facing page shows the relative dissonance (or consonance) of two complex tones, each of which consists of a fundamental frequency and six harmonic partials. Here we see peaks of consonance at the familiar musical intervals. The consonance of the octave is the easiest to explain.

If we designate 250 Hz as $f_0$, the partials of the lower note of the octave are $f_0$, $2f_0$, $3f_0$, $4f_0$, $5f_0$, and $6f_0$, and those of the upper note are $2f_0$, $4f_0$, $6f_0$, $8f_0$, $10f_0$, and $12f_0$; so the partials either coincide or are well-separated.

Next to the octave, the fifth is the most nearly consonant interval, traditionally and as shown in the graph at the top of the next page. The figure at the bottom indicates why this is so. In this figure horizontal distance is a measure of frequency separation in octaves; that is, frequency components with ratios of 2 to 1 are 1 octave apart. The six harmonies of the lower tone are shown as short vertical lines above the axis, labeled 1 through 6. The six harmonic partials of the upper tone, a fifth or .58 octave above the lower tone, are shown as short vertical lines below the horizontal lines, labeled 1 through 6. Note that two partials of the upper and lower notes coincide. Most others are separated by more than a quarter of an octave. The third partial of the upper tone falls between the fifth partial of the lower tone (.15 of an octave away) and the fourth partial of the lower tone (.17 of an octave away). Thus the smallest separation of any two partials when two tones (each with six harmonic partials) a fifth apart are sounded together is about a whole tone (.17 octave). This is only a semitone less than a minor third (.25 octave).

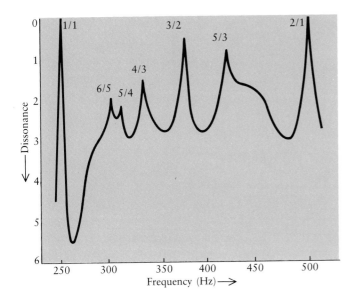

■ If we know the critical band-width and use the curve in the figure on page 77, we can calculate the consonance of a pair of tones, each of which has six harmonic partials, as a function of the frequency separation of their fundamentals. The curve in this figure is based on such a calculation. The frequency of the fundamental of the lower tone is fixed at 250 Hz, and that for the upper tone varies from a little below 250 Hz to a little above 500 Hz (an octave above 250 Hz). The ratios above the peaks of consonance show the frequency ratios of the traditionally consonant intervals: 1/1, unison; 6/5, minor third; 5/4, major third; 4/3, fourth; 3/2, fifth; 5/3, sixth; 2/1, octave.

We may note that the separation between the fifth and sixth partials of each tone is .26 octave, just greater than a critical bandwidth. If more partials are included in an individual tone, the higher partials will be less than a quarter of an octave apart, and the tone will have a sort of internal dissonance. We hear this as the buzzy quality of crude, electronically produced sounds, such as square waves and sawtooth waves. We also hear it in the jangly quality of the harpsichord. Of course, if notes are somewhat dissonant when sounded singly, they will be even more dissonant when sounded together.

We could make diagrams like that in the figure below for other intervals, such as the minor and major thirds, the fourth, and the sixth. In each case we would find some coinciding partials, and some partials closer (measured in fractions of an octave) than any in the diagram shown. But for dissonant intervals, such as a half tone, a whole tone, a tritone (an augmented fourth or six semitones), or a seventh, many partials would lie much closer together than a quarter octave; sounded together, notes separated by these intervals sound rough and dissonant.

■ The proximity of the partials for two tones, each with six harmonic partials, separated in frequency by the musical interval of a fifth. This interval is consonant because most of the partials of the two tones either coincide or are separated by more than a minor third. The third and sixth partials of the lower tone coincide with the second and fourth partials of the upper tone.

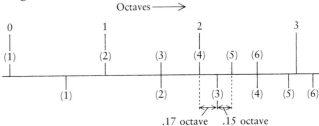

■ A cadence (left) and a deceptive cadence (right), which ends on a chord of the sixth (submediant) instead of on a tonic chord. The final chord of the deceptive cadence is consonant, but the progression sounds odd and not final.

The foregoing may seem to imply that we have found the root of consonance and harmony in the avoidance of partials that lie too close together in frequency. If partials are too close, they will beat or sound rough. For tones made up of many (say, six) harmonic partials, we can avoid serious beats or roughness by sounding together notes whose fundamental frequencies have integer ratios, such as 3/2 (fifth), 5/3 (sixth), 4/3 (fourth), 5/4 (major third), or 6/5 (minor third). Have we indeed found the unassailable basis of consonance? Perhaps.

Why the reservation? For a pair of sine waves, naive listeners will judge *any* interval greater than a critical bandwidth as consonant; this is true even for the seventh, and even for intervals with irrational ratios of frequencies. Trained musicians will dutifully identify thirds, fourths, fifths, and sixths, and report them as consonant, but will report sevenths as dissonant.

For complex tones made up of many partials, some musicians give strange judgments. I have heard one characterize an isolated deceptive cadence, shown at the right in the figure above, as dissonant. Both final chords in the cadence (left) and the deceptive cadence (right) are consonant, but the final chord in the deceptive cadence is unexpected and "wrong," as you can hear by playing it on the piano.

Much simple music closes with a cadence from the dominant seventh to the tonic (see the figure at the left). The dominant seventh is a dissonant chord. Is this why the cadence is effective and identifiable? Perhaps historically. Producing sounds by means of a computer, Max Mathews and I carefully deleted all partials closer than a quarter octave in both chords, rendering the dominant seventh chord consonant, and the tonic chord somewhat more consonant than is "natural." A trained musician identified the chords properly, even though the dominant seventh was no longer acoustically dissonant. The "clue" was most likely the half-tone step from F to E, in addition to the half-tone step from the leading tone (B) to C.

■ A cadence that goes from the dominant seventh, a dissonant chord, to the tonic, a consonant chord.

It is clear that how we "hear" musical sounds depends partly on musical training. It is also clear there is a basis for consonance and dissonance in the very nature of human hearing, since the critical bandwidth is related to the way that the auditory nerves transmit messages to the brain. Nevertheless, there is more to harmony than an avoidance of dissonance. We can see this from Helmholtz's remarks about the beginning of Palestrina's eight-part *Stabat Mater* (on the next page). Of this passage, which is certainly consonant, Helmholtz says,

■ This score is the beginning of *Stabat Mater* by Palestrina, cited by Helmholtz. The passage is consonant, but, to Helmholtz, the succession of "chords" seemed odd because they did not clearly establish a key. Perhaps Palestrina thought of the voices as consonant melodic lines rather than as chords.

Here, at the commencement of a piece, just where we should require a steady characterization of the key, we find a series of chords in the most varied keys, from A major to F major, apparently thrown together at haphazard, contrary to all rules of modulation. What person ignorant of ecclesiastical modes could guess the tonic of the piece from this commencement? As such we find D at the end of the first strophe, and the sharpening of C to C♯ in the first chord also points to D. The principal melody, too, which is given in the tenor, shows from the commencement that D is the tonic. But we do not get a minor chord of D till the eighth bar, whereas a modern composer would have been forced to introduce it in the first good place in the first bar.

I asked a modern composer, physicist, and acoustician about this. His first comment was that the passage could easily be interpreted as being in the key of D, but the next day he said that it depended on how one looked at it. A more learned friend, Gerald Strang, said, "To me it is clear Dorian mode."

We will explore the ideas of consonance and harmony further in Chapter 6.

# Rameau and Harmony

For many years I was convinced that the work of Helmholtz, as augmented by Plomp, had established the physical and psychophysical basis of both consonance and harmony. Things seemed so simple. Sounds were consonant when no (or few) frequency components (partials) fell within the same critical bandwidth. This concept explains both the traditionally consonant intervals, such as the octave and the fifth, and the traditionally dissonant intervals, such as the second or the seventh.

Traditionally, in counterpoint of any number of voices, a dissonance occurs only when any one voice "offends" against any other. Likewise in chords consonant by conventional standards of harmony, all the intervals among notes meet Helmholtz's criterion to some degree. Thus I understood harmony as an outgrowth of counterpoint.

Led on by the ideas of Helmholtz and Plomp, in 1966 I proposed to attain the effects of musical harmony by synthesizing an entirely new scale. This scale consisted of eight notes spanning the octave, as shown in part B of the figure on the next page. The frequency ratio between any two successive notes in this scale is 1.0905; so the frequency ratio between the first and the third notes of the scale is 1.1892, which is the ratio for an equal-tempered minor third.

In effect, I had used C, E♭, G♭, and A as notes of my scale, and then had added a note 1.5 semitones above each of these notes.

Not only did I use a new scale, but for it I synthesized new tones in which all partials except the octave partials were nonharmonic. In fact, the partials of each tone were simply the frequencies of every other higher note in the scale; so all partials were an (equal-tempered) minor third apart, as shown in parts C and D. To a musician, the tones sounded like diminished seventh chords. To me, they sounded rather fruity but not dissonant, in accordance with the theories considered in Chapter 5.

Which of these new tones are consonant when sounded together? Suppose that we distinguish the notes of my scale as *odd* and *even;* that is, the first, third, fifth, and seventh notes are odd, and the second, fourth, sixth, and eighth notes are even. The first six partials of an odd note are shown in part C of the figure and the first six partials of an even note are shown in part D. Successive odd notes are a quarter of an octave apart; so are successive even notes. This is the same interval as that between the partials in the tones that I synthesized. But

■ Frequencies in this illustration are measured in octaves, not in Hertz. Thus, the frequency of octave 2 is twice the frequency of octave 1, and the frequency of octave 3 is 4 times that of octave 1. The vertical bars along horizontal line **A** show the fundamental frequencies (first partials) of the notes of the equal-tempered chromatic (12-tone) scale. The vertical bars along horizontal line **B** show the frequencies of the first partials of the eight-tone scale that I used in the *Eight-Tone Canon*. The vertical bars along horizontal line **C** show the frequencies of the first six partials of an "odd" note, the first note in the eight-tone scale. The vertical bars along horizontal line **D** show the first six partials of an "even" note in the eight-tone scale. All partials of an even *or* of an odd note are one-fourth of an octave apart. Partials of odd notes coincide; partials of even notes coincide. But the partials of an even note and an odd note are only an eighth of an octave apart, a very dissonant interval.

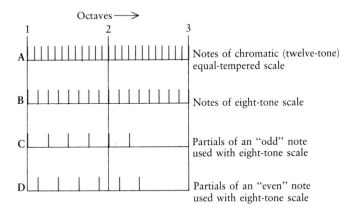

Octaves ⟶

A — Notes of chromatic (twelve-tone) equal-tempered scale

B — Notes of eight-tone scale

C — Partials of an "odd" note used with eight-tone scale

D — Partials of an "even" note used with eight-tone scale

adjacent odd and even notes in the scale are an eighth of an octave apart. An examination of the figure above and a little reflection will show that the laws of consonance for this scale and the tones used with it are very simple. When any two odd notes are sounded together, all partials of the two tones that are sounded coincide with one another (or miss completely). The same is true for any two even notes. Hence any pair of odd notes, or any pair of even notes, will be just as consonant as the tones themselves. However, if we sound an odd note and an even note together, some partials will lie an eighth of an octave apart. An eighth octave is less than a critical bandwidth; so odd notes sounded with even notes should sound very dissonant. Indeed, I found that they did.

Thus, for this eight-tone scale, the laws of consonance and, I thought, of harmony were extremely simple. Any and all odd notes go together; any and all even notes go together; odd and even notes sounded together result in dissonance.

To illustrate all this, I composed a four-voice *Eight-Tone Canon,* which was included in a Decca record, *The Voice of the Computer* (DL 71080), now out of print. In this canon I attempted to use transitions from dissonance to consonance in a "musically effective" way. I thought that the piece sounded pretty good, but in writing contrapuntally I had really evaded a clear test of harmony as we commonly understand it.

In 1979 I had an opportunity to spend a month in Paris at Boulez's IRCAM (Institute for Research and Coordination of Acoustics and Music). Emboldened

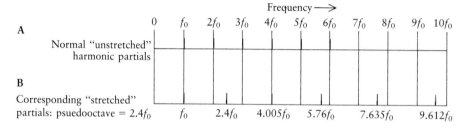

A

Normal "unstretched" harmonic partials

B

Corresponding "stretched" partials: psuedooctave = $2.4f_0$

Frequency $\longrightarrow$

| 0 | $f_0$ | $2f_0$ | $3f_0$ | $4f_0$ | $5f_0$ | $6f_0$ | $7f_0$ | $8f_0$ | $9f_0$ | $10f_0$ |

$f_0$ $\quad$ $2.4f_0$ $\quad$ $4.005f_0$ $\quad$ $5.76f_0$ $\quad$ $7.635f_0$ $\quad$ $9.612f_0$

<br>

86 | by the ideas of Helmholtz and my *Eight-Tone Canon*, I resolved to attempt a crucial experiment, one that Frank H. Slaymaker had performed in part in 1970. This experiment involved the playing of conventional chords and music with "stretched" octaves, or pseudooctaves, synthesized electronically. This concept of stretching is illustrated in the figure above.

Suppose that, instead of the octave, with a frequency ratio of 2, we use a pseudooctave whose ratio of frequencies is 2.4. Suppose that we also stretch, in a consistent way, the intervals between *all* frequency components of *all* partials of *all* notes that are sounded. A single tone in this system will no longer have harmonic partials (as in part A of the figure). Rather, it will have nonharmonic partials, whose frequencies are not integer multiples of the frequency of the first partial (as in part B). The successive (equal-tempered) semitones of the stretched chromatic scale will have frequency ratios of 1.0757 instead of 1.0594 as in the usual chromatic scale.

The result of this stretching of both the scale and the partials of the tones used with the scale is very simple. Suppose that we play the same music twice, first as *normal* (that is, unstretched and with harmonic partials) and second stretched and with stretched partials. If any two partials of two notes coincide in frequency in the unstretched version of the music, they will coincide also in the stretched version. Partials that don't coincide in frequency in the normal music will be a little farther apart, and hence a little more consonant, in the stretched version than in the normal version. Therefore, according to Helmholtz and Plomp, any combination of notes that is consonant in the normal music will be consonant, or a little more consonant, in the stretched music.

What did the stretched music sound like? We played a stretched version of *Old Hundredth* in four-part, note-against-note harmony for Pierre Boulez. He said that the only structure he could discern was the two fermata (double-length notes) at the middle and end of the piece. All familiar harmonic effects had vanished, and he apparently discerned no melody.

Those who heard the unstretched version first (we used both *Old Hundredth* and *The Coventry Carol*) could recognize the melody, and perhaps distinguish the lowest voice, whose first partial was the lowest frequency present. However, it was hard, if not impossible, to follow the inner voices. What of common and striking harmonic effects?

■ Frequency in Hertz is measured from left to right. The vertical bars along the upper horizontal line (**A**) show the frequencies of the harmonic partials of a "normal" tone whose fundamental is $f_0$. The vertical bars along the lower horizontal line (**B**) show the frequencies of the nonharmonic partials of a uniformly "stretched" tone in which the pseudooctave has a frequency of $2.4 f_0$ rather than the frequency $2 f_0$ of a true octave. Note that the spacing between successive harmonic partials is always the same but that between successive stretched bars increases with increasing frequency of the partials.

One of the most striking harmonic effects is the "finality" or "closing" quality of the *cadence,* a progression from the dominant chord (such as BDG) to the tonic chord (CEG), in which we go from the leading tone (the seventh, or B) to the tonic, C. Such a progression sounds like "The End" to anyone familiar with music. In the stretched harmony, musicians could not tell a cadence from an "anticadence" that went from the tonic to the dominant! Furthermore they could not tell a cadence from a deceptive cadence that ended on the chord of the sixth (CEA) rather than on the tonic. Neither seemed more final; both seemed equally strange and equally consonant (or dissonant).

Some work carried out at about the same time at Stanford by Elizabeth Cohen as a part of her doctoral work shed light on this. Among other things, she asked people to what degree tones with stretched partials seemed to be fused. That is, were stretched tones heard as a single sound, or simply as a collection of different frequency components? She found that, when the octave was stretched more than about 5 percent, the partials were not heard as a single tone of a particular timbre. Rather, listeners heard the collection of stretched partials as distinguishable tones of different frequencies (the higher-frequency partials were not individually distinct, but stood apart from the lower frequencies as a group).

This was something of a surprise to me, for many sounds with nonharmonic partials, such as the sounds of bells, gongs, drums, or knocking on wood, are heard as distinct, identifiable timbres. The ear does not analyze these sounds into separate components; yet that is just what our hearing did to the computer-produced tones with stretched partials. The ear simply refused to hear such a collection of partials as a distinct tone with a single pitch and a single distinctive quality. Instead, the ear picked these stretched tones to pieces.

This, I think, accounted for the confused impression made by four-part harmony with stretched chords. Because the ear could not identify the notes as single sounds, several notes sounded together were heard, not as a chord, either consonant or dissonant, but rather as a sort of mush of sound. That is why Boulez could hear no structure in a stretched *Old Hundredth,* even though all the mathematical structure present in the normal version was present in the stretched version, and, by the rules of Helmholtz and Plomp, consonance and dissonance were preserved.

In retrospect, I might have expected this result. When I visited the Philips labo-

■ Jan Schouten.

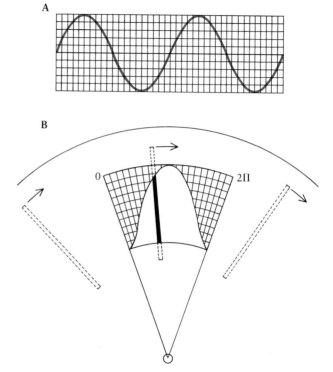

A

B

■ Diagram of Schouten's wave siren.

ratory in Eindhoven, Holland, shortly after World War II, Jan Schouten, an ingenious physicist, showed me a fascinating experiment. He had constructed a sort of optical siren by means of which he could produce sounds with various waveforms. Using this, he produced sounds with harmonically related partials of frequencies $f_0$, $2f_0$, $3f_0$, $4f_0$, and so forth. Then, by proper adjustments, he could remove the fundamental frequency $f_0$. I could hear this fundamental frequency come and go, but the pitch of the sound did not change at all. In some way, my ear inferred the proper pitch from the harmonics, each separated from the next by the frequency $f_0$ of the fundamental.

I later found that Schouten had published an account of this work in 1938, and in a paper published in 1940 he gave the name *residue pitch* to the pitch correctly heard in the absence of the fundamental frequency. This pitch has since come to be called *periodicity pitch*.

Schouten's observation should not surprise anyone who has listened to a pocket transistor radio. The speaker of such a radio is so small that the fundamental frequencies of all the lower notes are too weak to be audible; yet we hear the proper pitches, however tinny the music may sound. Later work has shown that we hear the pitch corresponding to the frequency difference between successive harmonics, even when several of the lower harmonics as well as the fundamental are missing.

A trained ear can to some degree analyze or pick apart a tone with harmonic partials, and hear separately the fundamental and various harmonics, at least up to the third harmonic, as Helmholtz observed. This was first demonstrated to me by D. E. Broadbent, an experimental psychologist who was then head of the Psychological Research Station in Cambridge, England. He successively added to a sine wave the second, third, fourth, fifth, and sixth harmonics. I heard the first three as tones of distinct pitches, but after the fourth harmonic was added, the individual components tended to merge, shortly after they were added, into a single sound with the pitch of the fundamental. (Experts assert that they can single out and hear partials separately up to the fifth or seventh.) Obviously, the ear has a strong tendency to ascribe a single pitch to a collection of tones whose frequencies are integer multiples of a common frequency, even though that frequency itself and some of its integer multiples are absent. Furthermore, we can have a sense of pitch and unity of sound even when the frequency intervals between successive higher partials are not exactly equal. This is how we ascribe a pitch to bells. This pitch is not the frequency of the lowest partial (the *hum tone*), but an average of the frequency separations between some higher partials of the sound of the bell.

In the tones with stretched partials, the frequency intervals between successive partials are not equal, as can be seen in part B of the figure on page 86. The frequency differences between these stretched partials are so different from the frequency of the first partial and from each other that the ear can ascribe no distinct pitch to, or hear any unity in, such a collection of partials; it therefore cannot fuse them into a single tone with a distinct timbre. If a single stretched tone is not heard as a distinct sound, can a chord made up of such tones have a distinct musical character? It seems that it cannot.

What does all this have to do with Rameau and his theory of harmony? Rameau attributed the distinct character of a major triad to a fundamental bass (*basse fundamentale*). What is this fundamental bass?

A major triad is a chord made up of a note of fundamental frequency $f_0$, the third above it, which has a fundamental frequency $(5/4)f_0$, and the fifth above it, which has a frequency $(3/2)f_0$; for example, CEG. As the figure on the next page shows, the harmonics of C will have frequencies

$$f_0, 2f_0, 3f_0, 4f_0, 5f_0, 6f_0,$$

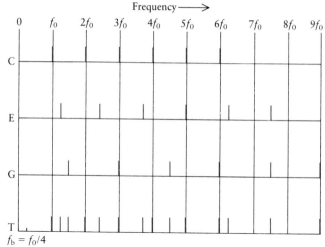

90 | the harmonics of E will have frequencies

$$(5/4)f_0, 2(5/4)f_0, 3(5/4)f_0, 4(5/4)f_0, 5(5/4)f_0, 6(5/4)f_0,$$

and the harmonics of G will have frequencies

$$(3/2)f_0, 2(3/2)f_0, 3(3/2)f_0, 4(3/2)f_0, 5(3/2)f_0, 6(3/2)f_0.$$

We see that all the partials of the notes present in the chord are integer multiples of the frequency

$$f_b = (1/4)f_0;$$

we can therefore write the harmonics of C as

$$4f_b, 8f_b, 12f_b, 16f_b, 20f_b, 24f_b,$$

the harmonics of E as

$$5f_b, 10f_b, 15f_b, 20f_b, 25f_b, 30f_b,$$

and the harmonics of G as

$$6f_b, 12f_b, 18f_b, 24f_b, 30f_b, 36f_b.$$

In the bottom row of the figure, the vertical bars indicate the frequencies of the first six partials of C, E, and G. Some of these coincide. Furthermore, many successive partials of $f_b$ are present in this triad. The frequencies $4f_b$, $5f_b$, and $6f_b$ are successive harmonics of $f_b$. So are $15f_b$ (the third harmonic of E) and $16f_b$ (the fourth harmonic of C), as well as $24f_b$ (the sixth harmonic of C) and $25f_b$ (the fifth harmonic of E). In listening to this triad, should we not therefore hear a pitch two octaves down from C, the root of the chord, corresponding to $f_b = f_0/4$? Experiments by Ernst Terhardt indicate that we do, and there is other evidence.

Rameau, too, must have heard this fundamental bass, two octaves below the root of the chord, when he listened to major chords. However, because he

■ In this figure, the vertical bars along the horizontal line C indicate the frequencies of the first six harmonic partials of C; those along line E show the frequencies of the first six partials of E; and those along line G show the first six partials of G. Together, these notes form a major triad. The frequencies of partials of all three notes (C, E, and G) are shown along line T. All the partials of all notes of the triad are integer multiples of a *fundamental bass* of frequency $f_b = f_0/4$. Rameau regarded an octave as an identity, and so he identified $f_0$, the fundamental of C, the root of the triad, as the fundamental bass of the chord.

regarded notes an octave apart as essentially identical, he called the root of the chord (C in our case) the fundamental bass, rather than the C two octaves below.

Rameau was the first to insist that what we now call various inversions of a chord are "the same chord" because they have the same fundamental bass. That is, EGC and GCE are the first and second inversions of CEG, and we regard them, as Rameau did, as essentially the same chord. Before Rameau, they were named differently and regarded as distinct chords. Rameau thus drastically reduced the number of "different" chords that a musician must learn.

If the distinct character of major chords lies in their having a fundamental bass, so that the partials of all notes present in the chord are integer multiples of this fundamental-bass frequency, then the lack of harmonic effect in stretched chords is easily understood. When we stretch a chord, the partials of the stretched tones of which it is composed are not integer multiples of any frequency whatsoever. In stretching the scale and the tones, we have destroyed the fundamental bass. The ear hears nothing to characterize the collection of stretched notes. Indeed, as we have seen, there is nothing that enables the ear to attribute a distinct pitch to the stretched tones of which we try to make up a stretched musical chord.

Although stretched tones seem vague collections of sinusoidal components rather than fused sounds with distinct pitches, if we play a stretched scale of stretched tones, we do get a sense of increasing pitch. Melody is more rugged than harmony. Melodies played with a stretched scale and stretched partials are easily recognized. It is only harmony that is confused or lost altogether.

There is a great deal of evidence that Helmholtz correctly explained the basis of consonance. As Plomp elaborated, when too many partials lie within a critical bandwidth, a sound will be rough or dissonant, whatever other qualities it may have. This concept adequately explains many things, including the fact that, although tunes played on a carillon sound all right, harmony on a carillon sounds discordant. When "consonant" chords are played on a carillon, many nonharmonic partials lie close together in frequency.

In Chapter 5 we noted that single notes as well as chords can have a dissonant character: the jangly sound of a note played on a harpsichord, or the buzzy

92 ■ Carillon of the church of
Notre-Dame at Anvers.

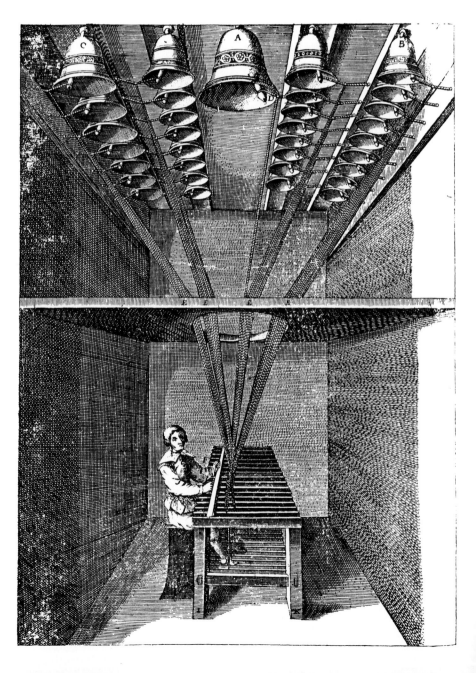

quality of an electronically generated sawtooth wave or square wave. But there is more to harmony than this sort of consonance and dissonance.

Indeed, there is more to harmony than major triads. Rameau himself had trouble explaining minor triads, which have no distinct fundamental bass. Indeed, early composers had trouble with minor keys, and often ended compositions with a "Picardy third," that is, used a major rather than a minor third in the tonic chord of the final cadence. Sometimes they entirely omitted the third in the final chord.

It may be that the major triad and its fundamental bass, which was first recognized by Rameau, is a sort of scaffolding on which very elaborate structures of harmony can be built. The major triads on the tonic, the dominant, and the subdominant (C, G, and F, for example) contain all the notes of the scale, as discussed earlier. Sounded in proximity, these three chords indicate the key unambiguously.

As the sole resource of harmony, these three chords would be dull. If we add a dissonant note, we can get the dominant seventh (GBDF), a most useful chord. The notes of a diminished seventh (BDFA♭) are the same in four different keys; so this chord, which has an ambiguous and unstable character, is useful for modulating from one of these keys to another. A few extra notes will make any major or minor chord sound lusher or jazzier, or may lend it a provocative touch of some other chord or some other key. Likewise, a chord can be rendered provocatively ambiguous by the omission of some of its notes.

To conclude this chapter, let me concede that nonharmonic partials are indeed important. Who would want to do without the distinctive and pleasing sounds of bells and gongs? Surely musical uses will be found for sounds that have entirely novel assortments of nonharmonic partials. But, to the brave people who seek such new sounds and new uses, I would say, judging from my experience, Beware; it isn't as easy as it looks.

# Ears to Hear With

**7**

All the wonderful sounds of music reach the brain through the mechanism of the ear and the nerves that connect the ear to the brain. A great deal is known about the structure of the ear and about the neural pathways from the ear to the brain, but our sense of hearing is understood only in part.

The upper left-hand drawing on the facing page shows the outer, middle, and inner ear. The visible outer ear, on the side of the head, is called the *pinna* (and the two ears are the *pinnae*). For many years, students of hearing thought that the pinna wasn't very important. In 1967 an independent scientist named Wayne Batteau showed that it is.

If you fold your ears over, or fill the convolutions with wax or modeling clay, and with your eyes closed listen to a nearby sound, such as the jingling of keys on a ring, you will find that you cannot judge the *height* of the sound, that is, tell whether such a sound is in front of you, above at an angle, or directly overhead. Because of our pinnae, the sensitivity of the ear to very high frequency sounds changes markedly with both the direction of the sound source and the frequency of the sound. Somehow, this enables us to judge the height of the sound source.

The pinna channels sound waves to the auditory canal, or *meatus*, which is about 2.7 centimeters (1 inch) long. At the inner end is the ear drum, or *tympanic membrane.* The auditory canal acts as a rather broad-band resonator, with a resonant frequency of about 2,700 Hz. Together with characteristics of the middle and inner ear, this broad resonance helps determine the frequency at which our hearing is most acute, which is about 3,400 Hz.

The eardrum divides the outer ear (the pinna and auditory canal) from the middle ear, which consists of three tiny bones. As the eardrum vibrates in response to sound waves, the bones convey these vibrations to a third portion of the ear, the inner ear. The upper right-hand drawing on the facing page shows the bones of the middle ear. Their names are the hammer (*malleus,* which looks more like a club), the anvil (*incus,* which looks rather like a tooth), and the stirrup (*stapes,* which really does look like a stirrup). These bones, which are flexibly connected together, convey the eardrum's vibration to a membrane covering an opening called the *oval window,* which separates the air-filled middle ear from the fluid-filled inner ear.

In the inner ear, the vibrations of the sounds that we hear are converted into electric impulses, which travel along nerve fibers to the brain by means of a

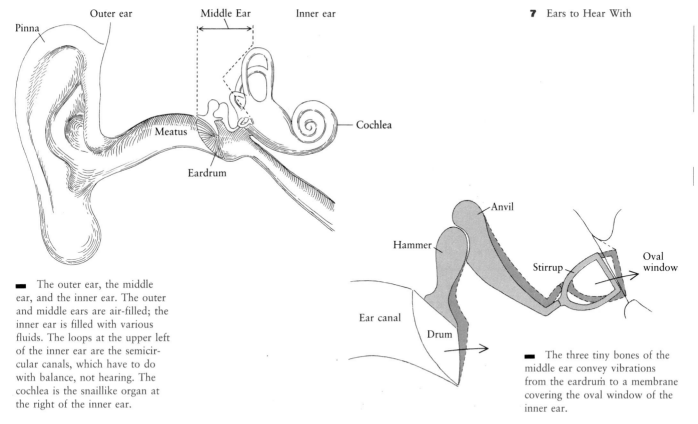

The outer ear, the middle ear, and the inner ear. The outer and middle ears are air-filled; the inner ear is filled with various fluids. The loops at the upper left of the inner ear are the semicircular canals, which have to do with balance, not hearing. The cochlea is the snaillike organ at the right of the inner ear.

The three tiny bones of the middle ear convey vibrations from the eardrum to a membrane covering the oval window of the inner ear.

Neural pathways from the hair cells in the cochlea to the cerebral cortex, a part of the surface of the brain. Pathways from both left and right ears are interconnected at various "way stations." A good deal is known about the pathways and their interconnections; less about the functioning of this complex system.

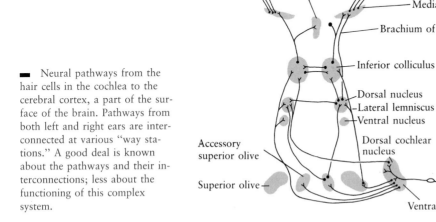

complicated, interlinked system of nerve pathways (see the figure above). Much is known about the structures of the neural pathways to the brain and the "way stations" at which they are interconnected. Less is known about the purposes they serve. Clearly, somewhere in the pathways, signals from the left and right ears are compared in such a way that we can tell the direction of a sound source.

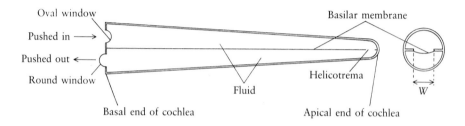

Oval window · Pushed in → · Pushed out ← · Round window · Basal end of cochlea · Fluid · Helicotrema · Apical end of cochlea · Basilar membrane · W

98 | The inner ear, enclosed in a bony case, contains three semicircular canals, which do not take part in our sense of hearing, but give us our sense of balance. The part of the inner ear that we are interested in is the *cochlea*, a spiral, tapered tube like the inside of a snail shell.

The cochlea is shown on the right in the upper left-hand figure on page 97. The spiral tube of the cochlea makes two and a half turns, and is about 3 cm (1.2 inches) long. It is about 0.9 cm (0.4 inch) in diameter at the beginning (basal) end, and about 0.3 cm (0.2 inch) in diameter at the far or apical end.

In order to be more easily understood, the cochlea is customarily drawn as a straight, tapered tube, that is, unrolled (as in the above figure). The schematic cross section of the cochlea is shown at the right in the figure. W indicates the width of the basilar membrane. This springy membrane is shown arched downward, as it would be if the pressure in the fluid above was greater than the pressure in the fluid below.

The width, W, and the stiffness of the basilar membrane vary from the left (basal) end to the right (apical) end. The membrane is widest and laxest at the right end, and narrowest and stiffest at the left end.

The longitudinal, or end-to-end, section of the unrolled cochlea is shown at the left in the figure. The basilar membrane does not extend all the way to the closed, apical end (on the right). Fluid above the basilar membrane can flow into the space under the membrane through an opening called the *helicotrema*.

At the basal end (at the left), there are two windows in the bone that surrounds the tube of the cochlea. These are covered with thin, flexible membranes. The upper opening is called the *oval window*. The stirrup (see the upper right-hand figure on page 97) is connected to the membrane covering the oval window. When a sound wave causes the stirrup to move in, it pushes the membrane in and causes the fluid in the upper part of the cochlea to move to the right; when the sound wave causes the stirrup to move out, it pulls the membrane out and causes the fluid in the upper part of the cochlea to move left.

The membrane that covers the round window below the basilar membrane isn't connected to anything. This membrane flexes in and out in accord with the pressure of the fluid under the basilar membrane.

If we pushed the oval window in slowly, fluid above the basilar membrane

■ The cochlea "unrolled" and shown as a straight tapered tube. The cross section (at the right) is simplified to a circle with rigid inner projections that support the springy basilar membrane, whose width increases along the length of the cochlea, from the basal to the apical end. The longitudinal (end to end) section of the unrolled cochlea is shown at the left.

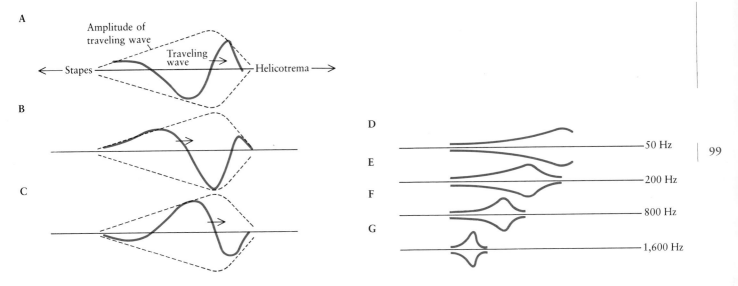

would flow to the right, pass through the helicotrema, flow to the left under the basilar membrane, and push the membrane over the round window out. The figure on the facing page shows the membrane over the oval window pushed in (by the stirrup) and the membrane over the round window pushed out by the motion of the fluid through the helicotrema. Actually, a sound wave causes the stirrup to move rapidly right and left, thus increasing and decreasing the pressure in the fluid above the basilar membrane. This causes a wave to travel along the membrane, through the cochlea from left to right. We can best visualize this wave as a traveling, up-and-down motion of the basilar membrane.

Let us assume that the stirrup moves left and right sinusoidally, with a frequency $f$. The speed with which the wave travels to the right along the basilar membrane depends on this frequency $f$. It also depends on the mass per unit length of the fluid above and below the membrane, on the mass per unit length of the membrane, and on the stiffness of the membrane. The cross section of the cochlea, the width of the basilar membrane, and its stiffness vary with distance along the cochlea; so the speed with which a wave of a given frequency travels along the basilar membrane decreases with distance from the left (basal) end. In fact, at some *place* along the cochlea the speed of the wave becomes zero. Near this place the basilar membrane oscillates up and down most strongly. At this place the wave stops, and its energy is absorbed.

The motion of a wave along the cochlea from left to right is shown in the left-hand part of the figure above. Views A, B, and C show the same wave as it moves to the right. The dashed lines show the envelope of the wave; that is, the greatest movement of the basilar membrane up or down at each point as the wave travels past that point.

Waves traveling along the basilar membrane from left to right: A, B, and C show the same wave as it moves toward the right; D, E, F, and G show the envelopes for waves of four different frequencies.

Parts D through G show the envelopes of waves along the basilar membrane for four different frequencies. We see that for low frequencies the place of greatest motion is closer to the apical end of the cochlea, for high frequencies the greatest motion is closer to the basal end. That is, the greatest vibration of the basilar membrane in response to a sinusoidal sound occurs at a particular place that depends on the frequency. The vibrations of the basilar membrane excite electric pulses in nerve fibers that end on hair cells at this place (see the figure below).

The reader may have seen by now that the frequency-dependent place at which the motion of the basilar membrane is greatest provides a means for sensing frequency and hence pitch. Sine waves of different frequencies send messages to the brain along different nerve fibers; so the brain may judge the pitch of a wave by "knowing" which particular fibers carry the message to the brain. This is called the *place theory* of pitch perception.

There is something more, however. A sharp pulse of sound (a click) sounds just the same whether the pulse is a brief increase in pressure that pushes the eardrum in, or a brief decrease in pressure that pulls the eardrum out. When pulses or clicks follow one another in rapid succession, the ear can sense the rate at which the pulses arrive (if it is not too fast); so we can also sense the periodicity or pitch from the rate at which pulses of sound arrive at the eardrum.

Sometimes the information conveyed by pulse rate conflicts with the frequency analysis performed in the cochlea. For example, the figure on the facing page shows two different trains of pulses. In Part A, all the pulses are of the same sign; that is, all represent an increase in air pressure. In part B, pulses are alter-

▬ The basilar membrane and the structures surrounding it. As the basilar membrane moves, the inner and outer hair cells send electric impulses along nerve fibers to the brain.

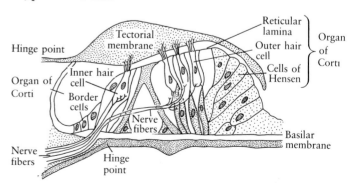

A

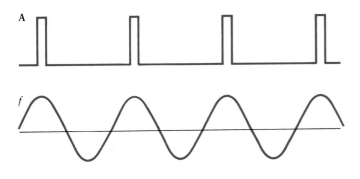

*f*

B

*f*/2

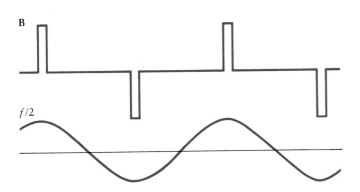

◼ Two different pulse trains of sound. In part **A**, successive pulses represent sharp increases of air pressure. In part **B**, alternate pulses are positive and negative, that is, represent sharp increases and decreases of air pressure. At 100 pulses per second, both pulse trains give the same pitch but, at 200 pulses per second, **B** sounds an octave lower than pulse train **A**.

nately positive and negative; that is, they alternately increase and decrease the air pressure. Individually, these tones are not very musical, and in some frequency ranges the tone in part B may lead to strange pitch judgments. But they do lend themselves to an experiment that sheds an interesting light on pitch perception.

When the frequency of the pulses of part A is low, say, 100 pulses per second, part B will give the same subjective pitch as part A when part B also has 100 pulses per second. But when part A has some higher pulse rate, say, 200 pulses per second, part B must have 400 pulses per second to match part A in pitch. This is in accord with the frequencies of the first partials, the sine waves shown below the pulse trains. At intermediate pulse rates, the pitch of part B is ambiguous. There is competition between the place judgment of pitch, based on the frequency of the first partial (and perhaps of other higher partials, which form a residue pitch) and the pulse repetition rate, which is conveyed to the brain along the nerve fibers that go from the cochlea to the brain.

The ear has two ways of judging pitch. One is by the place (or places) of excitation along the basilar membrane, corresponding to the frequencies of the various partials of the sound. The other is the periodicity with which bursts of electric pulses are sent to the brain along various nerve fibers from various parts of the cochlea. For the pulse train B of the above figure, the frequency of bursts wins out below 100 bursts per second; the place of excitation wins out above 100 bursts a second.

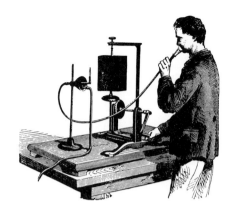

Actually, time information is preserved in the nerve pathways at much higher rates. Suppose that we send a sine wave of constant frequency to one ear, and send the other ear a sine wave of the same frequency but of different phase. The sound that we hear will seem to be inside the head, but closer to the ear in which the sine wave peaks first. This effect persists for frequencies as high as 1,000 to 1,500 Hz. When a sine wave of one frequency goes to one ear and a sine wave of a slightly different frequency goes to the other ear, most of us hear *binaural beats*. These persist up to frequencies of 1,300 to 1,500 Hz. Animal experiments in which electric nerve impulses are observed by means of very fine electrodes inserted into individual nerve fibers show that time information is present for frequencies up to 4,000 to 5,000 Hz, but it is not clear that animals use such high-frequency time information.

It is clear that both time and place (of maximum motion of the basilar membrane) are available to us in judging the pitch of periodic sounds. For nonperiodic sounds, we must rely on place information only. When we hiss by blowing air past our tongues, we can produce a sound of lower or higher pitch by moving the tongue back or forward. A hissing sound is not periodic, and this sense of pitch of a nonperiodic hiss must be judged by the place of maximum vibration along the basilar membrane.

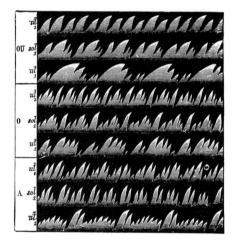

The place mechanism is also important in perceiving the timbre or quality of sounds. Suppose that in singing the same note we sing successively the vowel *u* as in *blue* and the vowel *e* as in *he*. We hear the pitch of each vowel as the same, and this pitch agrees with the frequency of the first partial of the note sung. However, *e* sounds shriller than *u*.

Indeed, we distinguish between vowel sounds largely by means of certain *formants* or frequency regions of high energy. These formants are centered on the resonant frequencies of the vocal tract, and are changed when we speak by changes in the shape of the tract. The frequencies of the formants all lie above the first partial of the sound we utter. The formants do not change frequency as we change the pitch with which we speak or sing a vowel. (This isn't quite true for sopranos; they raise the frequencies of formants in singing high notes.) We must perceive the formant frequencies, and hence distinguish among vowels, by means of the place mechanism of the ear.

It is by means of the place mechanism that we distinguish the various frequency

■ In a nineteenth-century experiment, it was found that vowel sounds would consistently impose the same pattern on the flame seen in the revolving mirrored drum.

components present in a sound. In theory, the critical bandwidth discussed in Chapter 6 should correspond to the "frequency range" along the basilar membrane that a single sine wave excites strongly. Another sine wave of a different frequency would lie in a different critical bandwidth if its excitation area along the basilar membrane doesn't much overlap that of the first. Do critical bandwidths and other phenomena of hearing agree quantitatively with the behavior of waves on the basilar membrane? We could answer this only if we knew exactly what motions of the basilar membrane are important in the excitation of the hair cells. Is it the displacement, the speed of motion, the curvature? And we would have to know the exact shapes of the envelopes of the motion along the cochlea shown in the figure of page 99.

Georg von Békésy, who received a Nobel Prize for his work on the ear and hearing, made experiments with cochleas excised from human cadavers. He found the envelopes of the displacement of the basilar membrane in response to a sine wave to be rather broad, as shown in the figure on page 99. Later experiments made with living animals indicated narrower envelopes, which broaden quickly after the animal dies. Some added neural "sharpening" mechanism may be needed to explain how such broad waves on the basilar membrane can give rise to the precision with which humans (and animals) can discriminate between sounds.

What we do know is that, in judging the pitch and quality of sounds, the ear supplies the brain with two sorts of information. One is information about time of occurrence and rate of repetition. The other is information about the frequency spectrum of a sound, which is derived from the intensity of vibration of the basilar membrane as a function of distance along that membrane. From these clues of time and place, we derive our sense of the pitches and qualities of sounds. By comparing the information coming from the two ears, we somehow find out what direction a sound is coming from.

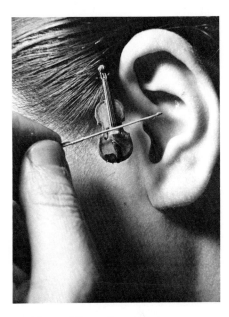

■ The world's smallest violin is quite shrill.

# Power and Loudness

**8**

We have all heard tales of the singer who shatters a glass with his (or her) voice. Such shattering is a purely physical phenomenon that depends on the power of the sound and the physical properties of the glass. Sounds can be very intense. They can shake a wall. They can shake us. Yet the most powerful sound is not loud to a deaf person.

Like pitch, loudness is something internal. For most musical sounds, pitch depends on just one property of the sound wave: its periodicity. The relation between the power of a sound and its loudness is much more complicated. However, loudness is obviously related to, and can be explained only in terms of, the power of a sound.

A sound source, such as a talker, a trombone, or a hi-fi speaker, radiates sound waves whose total power we can measure in watts, the same measure that we use for electric power. The sound waves spread out in all directions. The measure of the power of a sound wave that reaches our ears is the power density measured in watts per square meter. This is the *intensity* of the sound wave. Instead of specifying the power density itself, it is customary to specify how many times the actual power is greater than the power of a *reference-level* sound wave with a power of $10^{-12}$ watt per square meter\*. This reference level of power density is about the weakest sound that we can hear. The intensity of a sound measured in *decibels* above this reference level is the *intensity level*.

Most musical sounds have intensities that are millions of times the power of the reference level of sound intensity. Partly because it is difficult to write, read, or talk about huge numbers, and partly for other reasons, power ratios and intensity ratios are expressed in terms of decibels, abbreviated as dB. A certain number of decibels specifies a certain ratio of powers or intensities (power densities), as shown in the table on the facing page. The dB scale is logarithmic: a power ratio of 100 corresponds to 10 dB, and a power ratio of 10,000 to 40 dB; that is, the number of dB equals 10 times the number of zeroes in a multiple of 10.

In further discussions sound intensities are specified as a number of dB above reference level. *Sound level* as measured by a *sound-level meter* is measured in dB above reference level. About environmental noise, we speak of *noise level*.

---

\* Note that $10^{12}$ means one followed by 12 zeroes, but $10^{-12}$ means the reciprocal of this number, that is, one divided by $10^{12}$, or 0.000000000001.

Power ratios and decibels.

| $P/P_r$[a] | Ratio expressed in decibels[b] |
|---|---|
| 1,000,000 | 60 dB |
| 10,000 | 40 dB |
| 1,000 | 30 dB |
| 100 | 20 dB |
| 10 | 10 dB |
| 4 | 6 dB |
| 1 | 0 dB |
| 1/4 (=0.25) | −6 dB |
| 1/10 (=0.1) | −10 dB |
| 1/100 (=0.01) | −20 dB |

[a] $P$ is the power of the measured sound in watts per square meter; $P_r$ is the reference-level power.
[b] Mathematically, the number of decibels is $10 \log_{10}(P/P_r)$.

■ The intensity of sound or noise is measured with a sound-level meter. The meter does not ordinarily add the power of all frequency components of the sound and measure the sum. Rather, it "weights" the powers of different frequency components before adding. Weighting curve A is commonly used, because the reading so obtained is closely related to loudness to the human ear. Sound levels measured using the A setting (and weighting) of a sound-level meter are often quoted as dBA, as, a noise level of 40 dBA. To measure the actual power of a sound, we would use the C weighting.

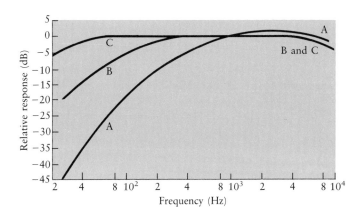

The reading of a sound-level meter doesn't give the total power of sounds of all frequencies. Sound waves of very low and extremely high frequencies we simply don't hear. As will be seen in this chapter, the ear is more sensitive to some frequencies than to others. The sound-level meter *weights* the powers or intensities of the frequency components of a sound wave before adding them to give an overall reading. The figure above shows three standard weighting curves.

Our ears are extremely sensitive. Suppose that we calculate the greatest distance at which we could just hear a 1-watt, 3,500-Hz sound source, the distance at which the sound will be at the threshold of our hearing, which is close to reference level. If we assume that the sound spreads out equally in all directions, the answer we get is 564 kilometers (352 miles). Of course, there is no way of doing such an experiment, because there is too much interference; but the calculation does give an idea of the remarkable sensitivity of our hearing.

Usually we can't hear sounds whose intensities are near reference level because the world around us is noisy. Early books tell us that the rustle of leaves in a gentle breeze produces a sound level of about 10 dB, that the sound level in a quiet garden in London can be as low as 20 dB, that in a quiet London street in evening with no traffic the sound level is about 30 dB, and that the night noises in a city are about 40 dB. I think that our world has grown noisier. The left-hand table on the facing page gives noise levels for a wide range of environments, and none is as quiet as some cited above. If you want quiet today you must go into a studio or concert hall. The average sound level in a concert hall with an attentive audience is about 40 dB, though the sound level in the hall when it is empty should not exceed 35 dB.

We should have such background sound levels in mind when we consider the intensity of musical sounds. An orchestra can produce a wide range of sound levels in a concert hall, from around 40 dB (the same as that of an attentive audience) to 100 dB (a million times as much power). For comparison, the sound level of conversational speech ranges from 40 to 70 dB (a range of 30 dB, or a thousand times).

What about individual musical instruments? The right-hand table on the facing page gives the peak sound powers in watts for various instruments. It also gives the sound level at three meters from each instrument in open space, calculated by assuming that the sound travels equally in all directions, with no reflections. In a room, the sound level at a distance of three meters would be appreciably higher because of reflections from the walls.

What about sound level and distance? If sound is not reflected or interrupted, the intensity drops 6 dB (that is, to a fourth of its value) every time we double the distance. If the sound level is 90 dB at 3 meters from the instrument, it will be 84 dB at 6 meters, 78 dB at 12 meters, and 72 dB at 24 meters.

Noise levels for various sources and locations.

| Source or description of noise | Noise level (dBA) |
|---|---|
| Threshold of pain | 130 |
| Hammer blows on a steel plate (2 ft) | 114 |
| Riveter (35 ft) | 97 |
| Factories and shops | 50–75 |
| Busy street traffic | 68 |
| Ordinary conversation (3 ft) | 65 |
| Railroad station | 55–65 |
| Airport terminal | 55–65 |
| Stadiums | 55 |
| Large office | 60–65 |
| Factory office | 60–63 |
| Large store | 50–60 |
| Medium store | 45–60 |
| Restaurant and dining rooms | 45–55 |
| Medium office | 45–55 |
| Automobile at 50 mph | 45–50 |
| Garage | 55 |
| Small store | 45–55 |
| Hotel | 42 |
| Apartment | 42 |
| Home in large city | 40 |
| Home in the country | 30 |
| Motion picture theater, empty | 25–35 |
| Auditorium, empty | 25–35 |
| Concert hall, empty | 25–35 |
| Church, empty | 30 |
| Classroom, empty | 30 |
| Broadcast studio, no audience | 20–25 |
| Television studio, no audience | 25–35 |
| Television studio, audience | 30–40 |
| Sound motion-picture stage | 20–35 |
| Recording studio | 20–30 |
| Average whisper | 15–20 |
| Quiet whisper (3 ft) | 10–15 |
| Threshold of hearing | 0–5 |

For full, add 5 to 15 dB (applies to Motion picture theater, Auditorium, Concert hall, Church, and Classroom)

Peak sound powers for various instruments, and their sound level at a distance of 3 meters in the open.

| Instrument | Peak power (watts) | Decibels above reference |
|---|---|---|
| Clarinet | 0.05 | 86 |
| Bass viol | 0.16 | 92 |
| Piano | 0.27 | 94 |
| Trumpet | 0.31 | 94 |
| Trombone | 6. | 107 |
| Bass drum | 25. | 113 |

As long as we stick to sound power or intensity, things are straightforward. We have to remember that sound level expresses the ratio of the measured intensity to a reference intensity that is near the threshold of audibility, and that the ratio of intensities is expressed in dB rather than as "times."

What about loudness? The best way to get to loudness is somewhat indirectly.

First, we should note that people differ in their hearing, as in other matters. Some have very acute hearing; some are quite deaf. The U.S. Public Health Service made an extensive survey of the *threshold of hearing* of many individual people; that is, of the level of sound at which they could just hear sinusoidal sounds of various frequencies. The results are at the top of the facing page.

In this figure, frequency of a steady tone is plotted left and right, and sound intensity in dB is plotted up and down. The flat top curve shows the level of feeling, the intensity level above which sound hurts. We see that this is about 120 dB for all frequencies.

Each lower curve is designated by a percent: 1%, 50%, 99%. The 1% curve means that 1 percent of the subjects tested could hear a sound of a particular frequency whose intensity lay above the curve. For example, at 1,000 Hz, 1% of all individuals in the group tested could hear a pure tone whose intensity level was above 3 dB.

Clearly, how loud a sound is depends on who is listening. A sound that is inaudible to one person may be 10 or 20 dB above the lowest audible level for another.

This figure tells us something else. Our ability to detect a sinusoidal sound, and sometimes the loudness of the sound, depends on its frequency, as well as on its intensity (power density).

All of us judge a particular sound as being louder if its intensity is increased. However, we can't discuss quantitatively the loudness of sounds for all people; so we must single out a group of persons, those with good or *acute* hearing, who are a minority among us.

For this group, loudness is related to intensity and frequency by the curves shown in the lower figure. The curve at the bottom is for the threshold of hearing, the intensity level at which a sound of a particular frequency can just be

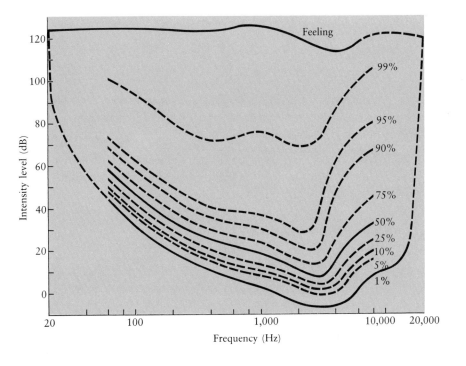

■ Curves showing threshold of hearing at various frequencies for a group of Americans: 1 percent of the group can hear any sound with an intensity above the 1 percent curve; 5 percent of the group can hear any sound with an intensity above the 5 percent curve; and so on.

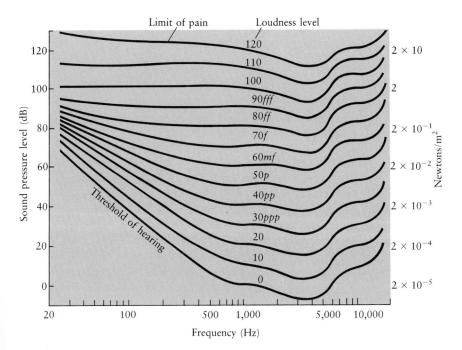

■ Constant-loudness curves for persons with acute hearing. All sinusoidal sounds whose levels lie on a single curve (an *isophon*) are equally loud. A particular loudness-level curve is designated as a loudness level of some number of *phons*. The number of phons is the sound level in decibels relative to the phon reference level, at which the curve crosses a vertical line at 1,000 Hz.

heard. The other curves are *equal-loudness curves* or *isophons*. That is, if two sinusoidal sounds of different frequencies have intensity levels lying on the same curve, they will sound equally loud.

We should note at once that the upper curves dip down less than the lower curves. For very loud sounds, the intensity level required to produce a given loudness doesn't change much with frequency. For very weak sounds it changes a lot. Equivalently, the loudness of a weak sound of a given intensity changes a lot as we vary its frequency; the loudness of a strong sound doesn't change much as we vary its frequency.

A consequence of this fact is that, as we turn the volume control of a hi-fi set, the *relative* loudnesses of sounds of various frequencies changes. Some hi-fi sets provide internal networks to compensate for this; mine has a switch labeled *loudness contour*. This is to be used when the sound produced by the speakers is less intense than the sound of the original music. When the loudness contour switch is on, the amplification is boosted at low and high frequencies to compensate for the dip in the low-intensity constant-loudness curves shown in the lower figure on the preceding page.

The equal-loudness curves in the figure are labeled in numbers of phons; 20 phons, 40 phons, and so on. A phon measures *loudness level*. The number of phons (of any particular equal-loudness contour) is merely the sound pressure level of an equally loud tone at 1,000 Hz. Any tone with the same loudness level in phons will have the same loudness, but what is loudness?

Loudness itself is measured in *sones*. A sound with a loudness of 20 sones sounds twice as loud as a sound with a loudness of 10 sones. A sound with a loudness of 50 sones sounds twice as loud as a sound with a loudness of 25 sones. In relating phons to sones, we simply ask a person to turn up the intensity level of a sinusoidal sound until he hears the sound as being twice as loud. Surprisingly, we get consistent results.

The figure at the top of the facing page shows the accepted relation between loudness in sones and loudness level in phons, which is simply the label of the equal-loudness curves in the lower figure on the preceding page.

We have noted that orchestral sound covers a range of intensity levels from 40 to 100 dB; this is approximately its loudness-level range in phons. This corre-

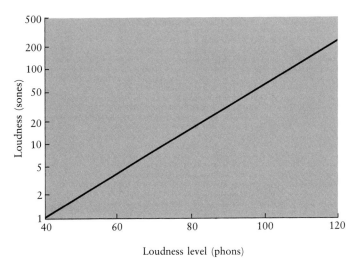

■ Relation between loudness in sones and loudness level in phons.

sponds to a range in sones from about 1 to 50, or to between five and six doublings of loudness.

We can detect very small changes in loudness, but this is largely irrelevant to music. The important thing is the number of loudnesses that we can notice or identify. Traditionally, the most common are *pianissimo (pp)*, *piano (p)*, *mezzopiano (mp)*, *mezzoforte (mf)*, *forte (f)*, and *fortissimo (ff)*. This amounts to six levels of loudness, which is about the same as the number of doublings of loudness between the weakest orchestral sound and the strongest (as in the table below). Is this coincidental? I don't think so. In music, the differences in dynamics or loudness that we specify correspond roughly to what a hearer regards as a doubling in loudness, that is, a doubling of loudness measured in sones; this accords with the markings from *ppp* to *fff* in the lower figure on page 111.

Sound intensities expressed in decibels.

|  | Intensity $(W/m^2)$ | Ratio $I/I_0$ | Level (dB) |
|---|---|---|---|
| Threshold of feeling | $10^0 = 1$ | $10^{12}$ | 120 |
| *fff* | $10^{-2}$ | $10^{10}$ | 100 |
| *f* | $10^{-4}$ | $10^8$ | 80 |
| *p* | $10^{-6}$ | $10^6$ | 60 |
| *ppp* | $10^{-8}$ | $10^4$ | 40 |
| Threshold of hearing | $10^{-12}$ | 1 | 0 |

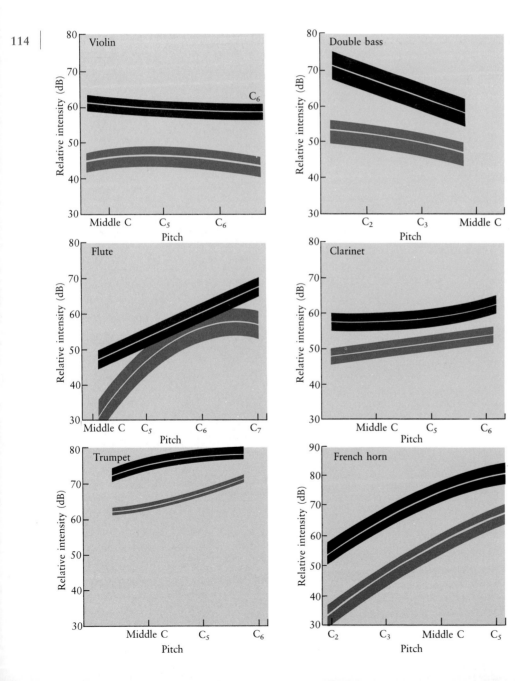

The dynamic ranges of various instruments. The broad upper bands show the relative intensities of the loudest tone that can be produced, as a function of pitch; the lower bands show the relative intensities of the softest tone that can be produced. The width of a band shows the range of intensities that may result when a player tries to produce "equally loud" tones.

The dynamic range of individual instruments other than the piano is much smaller than that of the orchestra, and for some wind instruments is exceedingly small. The figure on the facing page shows the relative intensities in dB of the loudest and softest tones that expert instrumentalists could produce at various pitches. One reason for orchestral music is to attain a wide range of sound intensity despite the limitations of individual instruments.

Are we done with loudness yet? Not quite. What of the loudness of two sounds sounded together? The rule is reasonably simple.

If two sounds of equal loudness are separated by more than a critical bandwidth,* when they are sounded together they are twice as loud as when either is sounded separately. More generally, the loudnesses of the sounds as measured in sones add.

What if several sounds lie within the same critical bandwidth? First we must add the *actual* intensities (not the intensity levels in dB) to get the total intensity. Then we must express the intensity level of this total intensity in dB above reference level. Then we must use the lower graph on page 111 to get the loudness level of the combined sounds in phons. Finally we must use the graph on page 113 to get the overall loudness in sones.

We then find that there is a great difference between sounds separated by more than a critical bandwidth and sounds separated by less than a critical bandwidth. For the former, to double the loudness we need add only two sounds of equal loudness. For sounds lying in the same critical band, we must add eight equally intense sounds in order to double the loudness.

Let us put this in another way. For simplicity, we will assume that the curves in the lower graph on page 111, which relate loudness level to sound intensity, are flat. For a single sine wave, we must multiply the power by 8 (an increase of 9 dB) to double the loudness. But we can double the loudness by adding two equally intense partials, separated by at least a critical bandwidth (say, harmonic partials of frequencies $f_0$ and $2f_0$). That is, for a sine wave we have to increase the power eightfold in order to double the loudness; for two well-separated frequencies we need merely double the power to double the loudness.

---

* About a quarter of an octave (a minor third); see Chapter 6.

We can put this another way. Suppose that some given power at a single frequency gives a particular loudness. We find that, if we divide this power between two frequencies, we increase the loudness by a factor of about 1.6, and, if we divide the power among the first six harmonic partials ($f_0$, $2f_0$, $3f_0$, $4f_0$, $5f_0$, $6f_0$), we increase the loudness by a factor of about 7.

Notoriously, sine waves don't sound very loud. A group of harmonic partials of the same intensity level (power density) sounds much louder, partly because below 3,500 Hz a higher-frequency sound of a given intensity is louder than a lower-frequency sound of the same intensity (see the lower figure on page 111). To get a loud sound, it pays to put the power at frequencies higher than the fundamental, and separated by more than a critical bandwidth.

Successive harmonic partials above the sixth lie within a critical bandwidth. The sound of a harpsichord has many very high partials, and many lie within one critical bandwidth. That is why they sound rough or jangly. The sound would be even more jangly if the individual loudnesses of these high partials added, but they don't. Because successive high partials lie within the same critical bandwidth, the slow rise of loudness with intensity shown in the graph on page 113 helps protect us from jangle.

We have met several unfamiliar terms in this chapter, and it may be useful to bring them all together in one place.

The total power of a source of sound is measured in watts, just as electric power is.

In a sound wave, it is the *power density* in watts per square meter that counts. This power density is called the *intensity* of the sound.

*Intensity level* expressed in dB (decibels) tells us the ratio of the intensity of a sound to the intensity of a sound of *reference level,* which has a power density of $10^{-12}$ watts per square meter. The intensity level of a sound can be measured with a *sound-level meter.* If we want the intensity level of a sinusoidal sound (pure tone) or of any instrumental sound to correspond to its actual power, we must set the sound-level meter to the flat or C scale, shown in the figure on page 107.

*Loudness level* is a label for a constant-loudness curve, one of the curves shown in the upper figure on page 111. As a label, we choose the intensity level of the curve at 1,000 Hz. The unit of loudness level is the *phon*. A sinusoidal sound of 1,000 Hz frequency and an intensity level of 70 dB has a loudness level of 70 phons. Any other sinusoidal sound whose intensity level lies on the same curve, that is, any sound of the same perceived loudness, has the same loudness level.

Sounds with the same loudness level have the same loudness, but the loudness level doesn't tell us the loudness directly. The unit of loudness is the *sone*. If we double the loudness measured in sones, we hear the loudness as doubling. We can get from loudness level in phons to loudness in sones by using the graph on page 113.

I have saved one popular term for last. This is *sensation level*. Sensation level is the number of dB above the lowest intensity level at which a particular sound can be heard. That is, sensation level is the number of dB above the threshold of hearing for a particular sound.

Suppose that an experimenter has a sound source with a volume control that is accurately calibrated in dB, but is not calibrated relative to some known sound level. For two different settings of the volume control, the experimenter knows the ratio of the intensities (that is, the difference between the sound levels expressed in dB), but he doesn't know the exact sound levels with respect to $10^{-12}$ watts per square meter. The experimenter can produce a sound of 60 dB sensation level by first setting the volume control to produce a sound that the subject can barely hear and then turning the volume control up 60 dB.

Specifying sound levels as sensation levels rather than as intensity levels or sound levels (with respect to $10^{-12}$ watts per square meter) is convenient for the experimenter. It is also sensible. The threshold of hearing varies with frequency. It is different for different people. For the same person, it is a little different at different times. For many acoustical phenomena, including masking, it seems more sensible to measure intensity relative to threshold than to some arbitrary intensity.

# Masking

**9**

We all know that weak sounds are drowned out by loud sounds. We might compare this to being unable to see in a blaze of extraneous light, but the ear is different from the eye. A bright light blinds us for a long time; the ear recovers very quickly. A bright light blinds us for all colors; a loud tone of a particular frequency renders sounds of only some frequencies inaudible.

In 1894, Alfred Mayer excoriated conductors for obliterating the sounds of violins with the deeper and more intense sounds of wind instruments. He observed that an intense sound of low pitch can mask a weaker sound of high pitch, but a sound of high pitch cannot mask a sound of low pitch. This fits well with our understanding of how waves travel along the basilar membrane, as described in Chapter 7. The place of greatest excitation of the basilar membrane for tones of low frequencies is toward the far or apical end of the cochlea, but that for tones of high frequency is toward the near or basal end of the cochlea. In traveling along the cochlea, the wave excited by a high-frequency tone will never reach the place of a low-frequency tone. But in reaching their place, the waves set up by low-frequency tones must travel past the places of all tones of higher frequency. We might expect that the excitation of the basilar membrane at these places could interfere with the perception of high-frequency tones. If the low-frequency tone is strong enough, it can indeed interfere with our hearing tones of higher frequency.

When a weak sound is obscured by a stronger sound, it is said to be *masked* by the stronger sound. The strong sound is called the *masker*. The weak sound that is masked is called the *maskee* or *signal*. Masking by a strong sound may be likened to an impairment of hearing. The masker, in effect, raises our threshold of hearing, that is, raises the intensity that a sound must have for us to just hear it.

The first systematic investigation of masking was carried out at Bell Laboratories in 1924 by R. L. Wegel and C. E. Lane, and the results are available in Harvey Fletcher's *Speech and Hearing in Communication*. The figure on the facing page shows these results.

The figure presents results for masking tones of six different frequencies. Let us consider one set of curves, that for a masker frequency of 1,200 Hz. The horizontal scale shows the frequency of the tone that is masked. The vertical scale shows *threshold shift* in dB, that is, the increase in intensity level of the maskee

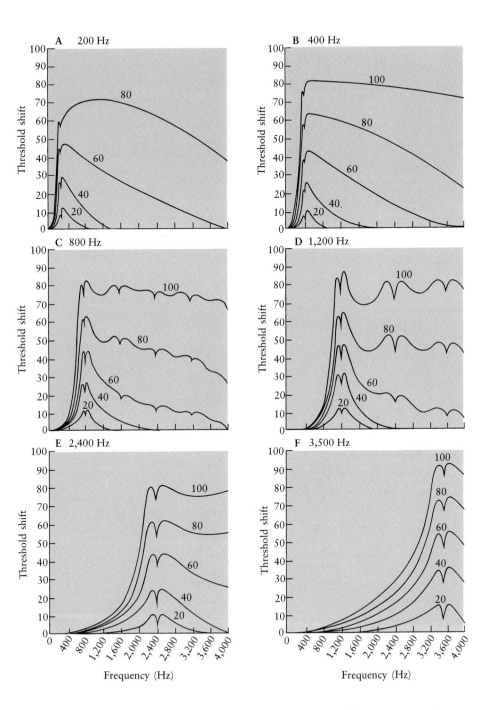

A   200 Hz

B   400 Hz

C   800 Hz

D   1,200 Hz

E   2,400 Hz

F   3,500 Hz

Frequency (Hz)

Threshold shift

■ Masking curves. In each box, the frequency of the tone that does the masking is given at the top; the horizontal scale gives the frequency of the tone that is masked; and the vertical scale gives the amount of masking in decibels, that is, how much more intense than threshold (in absence of masking) the masked tone must be in order to be heard in the presence of the masker. The marks above the curves are sensation levels for the masker, that is, how many decibels the level of the masker is above that at which it would just be heard. For a given curve (level of masker), the maskee will be heard if its sensation level lies above that curve, and will not be heard if its sensation level lies below the curve.

(above threshold in the absence of the masker) that will just enable us to hear the maskee in the presence of the masker. Five curves are shown, for masker sensation levels of 20, 40, 60, 80, and 100 dB.

Consider the topmost curve, for a very loud masker with a sensation level of 100 dB. A maskee will be heard if its sensation level above threshold *in the absence of the masker* lies above this curve. Consider two maskees of the same frequency, 1,600 Hz, one with sensation level (in the absence of the masker) of 80 dB, the other of 60 dB. The first maskee lies above the curve and will be heard; the second lies below the curve and won't be heard. In effect, the 100-dB masker has raised the threshold of hearing for 1,600-Hz sounds by about 69 dB above the threshold in the absence of the masker.

If we examine the curves, we see that there is no threshold shift for maskees with frequencies below 400 Hz. No matter how strong the 1,200-Hz masker, it causes no appreciable vibration of the basilar membrane at the 400-Hz place.

We note that the greater the intensity of the masker, the greater the upward shift of the threshold. We should expect this. As the intensity of a masker is steadily increased, a maskee of constant intensity must finally be masked.

We can also note that 1,200-Hz maskers with sensation levels of 20 and 40 dB produce no masking at frequencies above 2,400 Hz. Though the waves excited on the basilar membrane by such weak maskers pass the 2,400-Hz place on their way to the 1,200-Hz place, they don't interfere with the perception of 2,400-Hz tones.

For maskers of higher sensation levels, this is not so, and the shape of the curves for these higher-level maskers tells us something. Let us look at the 100-dB curve in part D, for example. We see little sharp dips at frequencies of 1,200 Hz, 2,400 Hz, and 3,600 Hz. *Very close to particular frequencies, the masker produces less masking than it does at nearby frequencies.* Why is this?

Consider masking by a 1,200-Hz masker of a tone of 1,220 Hz. We hear 20-Hz beats between the two tones. In essence, the combined amplitude of the sinusoidal tones of slightly different frequencies rises and falls with time. Sometimes the pressures of the masker and the maskee peak at the same time; a little later the pressure of the maskee is least when the pressure of the masker is greatest, and the combined pressure is lower than average. Still later the pressures add again. This rise and fall of the sum of the pressures, this beat between the two tones, repeats regularly at a rate of 20 Hz, the difference in frequency between the masker and the maskee. This beating enables us to detect the presence of a weak maskee even when we can't hear it as a tone of different frequency. Hence beats effectively lower the threshold shift and decrease the amount of masking.

This explains why there is a dip in the masking curve at 1,200 Hz, the frequency of the masker, but why are there dips at 2,400 and 3,600 Hz? Nonlinearities in either the electronic equipment used or in the middle or inner ear could cause a very strong 1,200-Hz masker to produce frequency components whose frequencies are harmonics of the 1,200-Hz masker, such as a 2,400-Hz second harmonic and a 3,600-Hz third harmonic. These waves of harmonic frequency would produce vibrations of the basilar membrane at the 2,400-Hz place and the 3,600-Hz place; these vibrations would beat with the maskee tones of frequencies near 2,400 Hz and 3,600 Hz and lower the masking curve.

Nonlinearities in the middle and inner ear could account for the fact that very intense maskers produce large threshold shifts for sounds whose frequencies lie above the frequency of the masker. As a wave of very large amplitude and low frequency, and waves of its harmonic frequencies, reached higher-frequency places on the basilar membrane, they could interfere with our ability to hear weak, high-frequency sounds.

If we look at the masking curves for low-level maskers, we find that masking extends for only a narrow range of frequencies. We might expect this frequency range to correspond to the critical bandwidth. In fact, the critical bandwidth for masking is pretty much the same as the critical bandwidth for dissonance and the critical bandwidth governing loudness.

So far we have discussed masking by single-frequency sinusoidal sounds, that is, by pure tones. Noise made up of a broad band of frequencies also produces masking. A colleague and I once shared an office that looked out on the West Side Highway in lower Manhattan. In summer, when the windows were open, streams of passing cars flooded the office with a broad-band noise, an intense sound with a continuous range of frequencies. I soon got used to the noise, and found that it gave me complete privacy. I couldn't hear a visitor talking to my colleague who sat at a desk behind me. I couldn't even hear his visitor enter or leave the room.

Many masking experiments are carried out using noise rather than pure tones. The figure below compares, for the same subject, masking by a pure tone of 400 Hz (part A) and masking by a band of noise extending from 365 Hz to 455 Hz (part B). We should note that in this figure the levels of the curve are *intensity levels* with (respect to $10^{-12}$ watt per square meter), not sensation levels as shown in the figure on page 121. We see from the lower figure on page 111 (Chapter 8) that at 400 Hz the threshold of hearing is at an intensity level of 10 dB; so the curve marked 60 in the figure below would correspond to a curve marked 50 dB (sensation level) in the figure on page 121.

In comparing masking by noise (part B) with masking by a pure tone (part A), we find that in masking by noise there are no dips due to beats. We hear noise plus a pure tone simply as noise plus a pure tone; we don't hear any beats.

The figure shows that, at most frequencies and levels, noise masks more effectively than does a pure tone of the same intensity. But the masking of a pure tone of intensity level 80 dB falls off less rapidly with increasing frequency than does masking by a narrow-band noise of the same intensity level.

■ Masking for a pure-tone masker (A) compared with masking for a 90-Hz band of noise (B). In part A, the masker frequency is 400 Hz; in part B, the masker (noise) is centered on 410 Hz. In these curves the labels on the curves specify sound level. The threshold of hearing at 400 Hz is about 10-dB intensity level; so the sensation level of the maskers for these curves is about 10 dB less than the intensity level, the labels shown on the curves. Narrow-band noise usually masks more than a pure tone, but for the highest curves the pure tone masks more at higher frequencies.

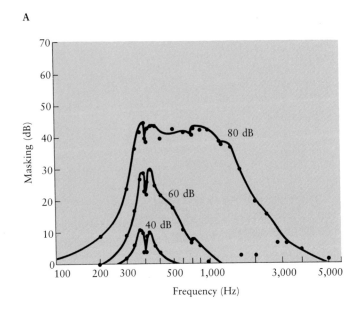

A

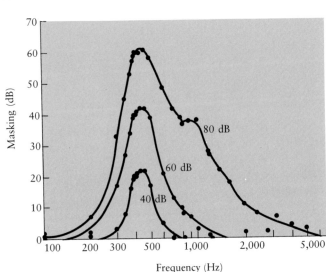

B

What about masking by wide-band noise? The only portion of the power of wide-band noise that is effective in masking a pure tone is the power of those frequency components whose frequencies lie within a critical bandwidth centered on the pure tone. The *power density* of *white noise* is independent of frequency; that is why the noise is called white. That is, the power or intensity of the part of the noise that lies within any band of frequencies is proportional to the bandwidth. If the critical bandwidth is proportional to frequency (say, a quarter octave), the effective intensity of the white noise in masking a pure tone will be proportional to the frequency of the tone.

Masking by white noise is important in psychoacoustic studies. In the perception of music, masking by other sounds is much more important. We noted in Chapter 8 that the intensity level in an auditorium with an attentive audience is around 40 dB. What does this mean to listeners? What level of musical sound will such audience noise mask? How can we find out? We don't have masking data for audience noise and musical instruments or orchestras. We must infer what we can from masking data for pure tones and for noise.

The figure on page 121 gives data for masking of a tone by a tone. We see from the figure that, for low masking levels, we can hear tones some 10 to 15 dB below the masker level. But audience noise is not a pure tone.

If we consult part B of the figure on the facing page, which gives masking of a tone by noise, we conclude that we can just hear a tone if it is some 7 dB below the level of *narrow-band* masking noise. (Remember that for this figure a masker level of 40 dB corresponds to a sensation level of 30 dB.) Other data on masking of tones by noise indicate that a tone becomes inaudible at from 2 to 6 dB below the noise in a critical bandwidth.

We have discussed the masking of a tone by a tone, and a tone by noise. Masking of narrow-band noise by a tone is very different. We tend to hear the noise despite the presence of the tone. In effect, the noise tends to make the tone wavering and noisy. Indeed, a critical bandwidth of noise can be masked completely by a tone in the center of the band only when the tone is about 24 dB stronger than the noise.

We conclude that, if musical sound had the same spectrum as audience noise, we could hear it if its level were a few dB below that of audience noise, perhaps in

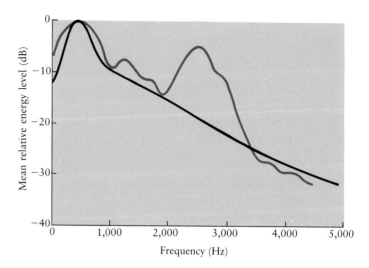

Comparison of the frequency distribution of sound power for an orchestra (shown in black) and for a tenor (shown in color). The curves are adjusted to have the same peak values. Orchestral sound falls off steadily in power as the frequency is increased above 500 Hz. The singer's voice has a second peak of power between 2,000 and 3,000 Hz, which gives him an advantage against masking of about 13 dB (or 20 times) in this frequency range.

the range from 34 to 38 dB for an audience-noise level of 40 dB. But we could hear a pure tone with a much lower intensity level, for only that part of the audience noise within a critical bandwidth around the pure tone would be effective in masking it. Saying that the noise produced by an attentive audience has an intensity level of 40 dB may give the false impression that such audience noise is quite annoying. This amount of noise won't much interfere either with many instruments playing together or with single instruments played softly.

Perhaps it is more pertinent to ask how singers manage to be heard above an orchestra. Johann Sundberg has found the explanation by comparing the frequency distribution of power (the sound power or intensity spectrum) for an orchestra and for a tenor (in this case, the late Jussi Björling). The figure above shows what he found. The intensity level of orchestral sound falls off rapidly at frequencies above 500 Hz. The power of the singer's voice peaks at about 500 Hz, but it peaks again at about 2,500 Hz, which gives the singer an advantage over the orchestra of about 13 dB (or 20 times in intensity). In this frequency range, the singer can equal the orchestra even if he produces only a twentieth as much sound power.

This discussion of masking is not quite finished. Have you ever noticed that in a crowded room you can hear what someone nearby is saying, despite the babble of other speakers who are no farther away? This ability to distinguish one voice among many is called the *cocktail-party effect*. It can be effective in sorting out the instrumental sounds of a chamber group if they play near one.

We might think that the cocktail-party effect is merely a matter of "paying attention." No doubt paying attention is important, but there is more to the cocktail-party effect than that. The effect vanishes if we listen over a single audio channel. In Chapter 7, it was noted that time information is preserved in the auditory pathways. Imagine that the speaker that we wish to hear is dead ahead, whereas the noise that we wish to ignore comes from the side. The sound of the speaker will reach both ears at the same time and in phase; the noise won't. We can take advantage of this in listening, so that somehow the neural speech signals add in phase and reinforce one another, whereas the noise signals don't add in phase and may partially cancel. (The mechanism of the cocktail-party effect is really more complex than this, and is not fully understood. It enables us to single out a voice coming from a direction other than dead ahead.)

The cocktail-party effect is extremely important in listening to *nearby* sounds, and it may help somewhat in ignoring the coughs of near neighbors while listening to music in a large hall. But, for concert audiences the phenomenon of masking, without regard to binaural effects, is predominant. It is also very important for distortion in sound systems used for the reproduction of music. Nonlinearity in sound systems produces frequency components that were not present in the original music. If these lie close to frequencies already present in the music, they will be masked and we won't notice them. But, if they fall in frequency ranges where the intensity of the music itself is small, the frequencies produced by nonlinearities can be very annoying. We notice this particularly in small rattles or in the noise produced when a misaligned speaker coil touches the magnet around it and produces a faint, high-frequency rustling sound as it moves. The power of this rustling sound is very small compared with the power of the music itself, but we hear the rustling sound clearly because it is not masked by the lower-frequency sounds of the music.

Like pitch and loudness, masking is of great importance in the perception of musical sound. Masking is important in the reproduction of sound, for unmasked distortion grates on the ear; masked distortion does not. Masking is important in listening to sound. In a noisy automobile we can't hear the soft passages in a composition; in a quiet concert hall we can.

Perhaps the most fundamental importance of the masking of musical sounds lies in the fact that one musical sound can mask another. Intended subtleties will be obscurities if the sounds intended to produce them are masked by other sounds.

# Other Phenomena of Hearing

Why do psychoacousticians single out and measure various phenomena of hearing? Partly, to find measurements that give consistent results. Partly, to accumulate data that will suggest, validate, or destroy various theories of hearing.

Many of the measurements made by psychoacousticians are of little practical value for music. For example, as we have seen, most musical tones have two quite different qualities: pitch, which depends on periodicity; and brightness, which depends on the relative strengths of the partials. Sine waves have only one frequency component, and in experiments done with sine waves, the naive listener might respond to either his sense of brightness or his sense of pitch. However, experimental data are our only guard against hasty conclusions and self-deceit. Although we would prefer to have more experimental data truly relevant to musical sounds, we should certainly not disregard what psychoacoustic data we do have.

One classical measurement of psychoacoustics is the *jnd* (just noticeable difference), or *limen*. The most common jnds are those of intensity and of frequency, measured for sine waves. Unhappily, measured jnds differ greatly, from one method of measurement to another, from one subject to another, and even for measurements taken on the same subject at different times.

In measuring jnds, some early experimenters "wobbled" the intensity or frequency, and found how small a wobble was just noticeable. An alternative is to present in succession two tones, *A* and *X* (for unknown), with slightly different intensities or frequencies, and to ask the subject whether the intensity or frequency of *X* is greater or less than that of *A*.

The jnd of intensity depends on both frequency and intensity. The left-hand table on the facing page shows a classic measurement of the jnd of intensity in dB for sine waves of various frequencies and sensation levels. The jnd is least, about a quarter of a dB, at high levels and at frequencies from 1,000 to 4,000 Hz. At frequencies above 100 Hz and intensity levels above 40 dB, it does not exceed 1 dB.

The jnd of frequency also depends on frequency and intensity. The right-hand table shows the jnd of frequency for pure tones of various frequencies and sensation levels. The jnd is least, about 3 cents, for a frequency of 2,000 Hz and for sensation levels of 30 dB or more. Above 100 Hz and 30-dB sensation level, it does not exceed 50 cents.

■ The photograph on the preceding pages is of Manfred Schroeder in the anechoic chamber at Bell Laboratories.

The minimum detectable changes (jnd) of intensity in decibels for sine waves.

| Frequency (Hz) | Sensation level | | | | | | | | | | | |
|---|---|---|---|---|---|---|---|---|---|---|---|---|
| | 5 | 10 | 20 | 30 | 40 | 50 | 60 | 70 | 80 | 90 | 100 | 110 |
| 35 | 9.3 | 7.8 | 4.3 | 1.8 | 1.8 | | | | | | | |
| 70 | 5.7 | 4.2 | 2.4 | 1.5 | 1.0 | .75 | .61 | .57 | | | | |
| 200 | 4.7 | 3.4 | 1.2 | 1.2 | .86 | .68 | .53 | .45 | .41 | .41 | | |
| 1,000 | 3.0 | 2.3 | 1.5 | 1.0 | .72 | .53 | .41 | .33 | .29 | .29 | .25 | .25 |
| 4,000 | 2.5 | 1.7 | 0.97 | 0.68 | .49 | .41 | .29 | .25 | .25 | .21 | .21 | |
| 8,000 | 4.0 | 2.8 | 1.5 | .9 | .68 | .61 | .53 | .49 | .45 | .41 | | |
| 10,000 | 4.7 | 3.3 | 1.7 | 1.1 | .86 | .75 | .68 | .61 | .57 | | | |

The minimum detectable changes (jnd) of frequency in cents for sine waves.

| Frequency (Hz) | Sensation level | | | | | | | | | | |
|---|---|---|---|---|---|---|---|---|---|---|---|
| | 5 | 10 | 15 | 20 | 30 | 40 | 50 | 60 | 70 | 80 | 90 |
| 31 | 220 | 150 | 120 | 97 | 76 | 70 | | | | | |
| 62 | 120 | 120 | 94 | 85 | 80 | 74 | 61 | 60 | | | |
| 125 | 100 | 73 | 57 | 52 | 46 | 43 | 48 | 47 | | | |
| 250 | 61 | 37 | 27 | 22 | 19 | 18 | 17 | 17 | 17 | 17 | |
| 500 | 28 | 19 | 14 | 12 | 10 | 9 | 7 | 6 | 7 | | |
| 1,000 | 16 | 11 | 8 | 7 | 6 | 6 | 6 | 6 | 5 | 5 | 4 |
| 2,000 | 14 | 6 | 5 | 4 | 3 | 3 | 3 | 3 | 3 | 3 | |
| 4,000 | 10 | 8 | 7 | 5 | 5 | 4 | 4 | 4 | 4 | | |
| 8,000 | 11 | 9 | 8 | 7 | 6 | 5 | 4 | 4 | | | |
| 11,700 | 12 | 10 | 7 | 6 | 6 | 6 | 5 | | | | |

Such measurements of just noticeable differences of intensity and frequency are not necessarily relevant to the performance of music. The measurements are made for sine waves, which aren't used in music. Performers asked to play an "even" scale deviate in intensity by about 5 dB, which is much greater than the jnd of intensity. It was noted in Chapter 8 that, if the degrees of loudness from *pp* to *ff* span a range of 60 dB, each is separated from the next by 12 dB. But most single instruments do not span a 60-dB range of intensity; so the deviations that performers make in sound intensity are large compared not only with jnds of intensity, but also with the range of intensity that such instruments can produce.

The jnd for frequency seems more relevant for performance. Early measurements indicated that, when pairs of pitches or musical intervals were played successively (melodically) or simultaneously (harmonically) by players of a string trio, the frequencies were more than 18 cents away from the true frequency at least half the time, and the greatest differences in intervals as played exceeded 40 cents. Later measurements have indicated deviations in frequency of about 10 cents for good violinists, a figure not too far from the jnds in the figure on page 131.

Perhaps jnds of frequency tell us how accurately a musician can tune an instrument. There was an electronic tuning test at the Bell System exhibit in Disneyland. By pressing buttons you could hear either a tone of fixed frequency or a tone whose frequency you could adjust, but not both at the same time (so you couldn't hear beats). After you had matched the frequencies as closely as possible, the machine scored your performance. My wife, who is a musician, did much better than I.

Other aspects of auditory perception are much more important to musical sound than are jnds of frequency or intensity. At the head of these I would put what is called the *precedence effect,* or the *Haas effect.*

This effect you can demonstrate to yourself if you have a stereo system, and especially if you tune to a monophonic (nonstereo) channel, so that exactly the same sound comes from each speaker. If you stand equidistant from both speakers, you hear the sound as coming from a phantom source midway between the speakers. But if you stand a foot or more closer to one speaker than to the other, all the sound seems to come from the nearer speaker.

I checked this recently in a car that had a stereo radio. In the exact middle of the front seat, I heard the announcer's voice as a compact sound source dead ahead, midway between the speakers. As I moved to the right the sound source at first became diffuse. As I moved farther, all the sound clearly came from the right-hand speaker.

Like other stereo systems, the car radio has a knob (a balance control) that changes the relative intensities of the sounds coming from the right and left speakers. When one is sitting in the driver's seat, for a monaural or centered signal (the announcer's voice, for example) all the sound seems to come from the left speaker unless the sound from the right speaker is made more intense than that from the left. Indeed, even for a stereo signal most of the sound seems to come from the left speaker unless the sound from the right speaker is made more intense. By making the sound from the right speaker more intense, we can hear sound from both speakers, and get a stereo effect. However, adjusting the relative intensities doesn't cure the disparity between time of arrival (at our ears) from the left and right speakers. If the speakers are equidistant and the intensities are equal, we hear a speaker's voice as coming from a distinct direction, from a compact source. If the speakers aren't equidistant, and we increase the intensity of sound from the far speaker, we do hear some sound from it, but we no longer hear a speaker's voice as coming from a compact source. The source of the voice seems extended and fuzzy.

I have said that, if a sound reaches us with equal intensities from two sources, we hear all of it as coming from the nearer source if the difference in distance is about a foot or greater. There's a limit to this, of course. If the difference between the distances to the two speakers is great enough (about 20 to 25 meters, or 60 to 80 feet), you hear two sound sources. The speaker that is farther away produces an echo of the sound from the nearer speaker. You can easily demonstrate this transition from fusion to echo as the American physicist Joseph Henry did about 1849. Stand in front of an extensive, smooth wall, and clap your hands. If you're less than ten meters from the wall (sound then travels 20 meters in reaching the wall and returning), you hear a single sound. If you're farther than 13 meters or 40 feet away, you hear an echo of your handclap.

The precedence effect, the fact that a sound seems to come from the direction from which it reaches us first, is bad for stereo, but highly desirable in everyday life. When someone speaks to you in a hard-walled room, you hear all the sound as coming from his or her mouth, even though much of the sound that reaches you has been reflected from the walls, sometimes several times. This reflected sound adds to the loudness, but doesn't keep you from identifying the direction of the source. The same is true for musical instruments. If you are close enough to a chamber group, you hear each sound coming from its proper direction, even though much of the total sound that you hear reaches you after reflection from the walls.

Reflections from the walls of a room do not confuse our sense of direction, but they add to the intensity of the sound, and they add to its quality as well. While I was in Paris, working at IRCAM, my wife practiced there on a Yamaha concert grand in a studio that had very sound-absorbing walls. She found this exasperating, because, loud as she tried to play, she produced very little sound.

Absorption coefficients of some building materials.

| Material | Frequency (Hz) | | | | | |
|---|---|---|---|---|---|---|
| | 125 | 250 | 500 | 1,000 | 2,000 | 4,000 |
| Marble or glazed tile | .01 | .01 | .01 | .01 | .02 | .02 |
| Concrete, unpainted | .01 | .01 | .01 | .02 | .02 | .03 |
| Asphalt tile on concrete | .02 | .03 | .03 | .03 | .03 | .02 |
| Heavy carpets on concrete | .02 | .06 | .14 | .37 | .60 | .65 |
| Heavey carpets on felt | .08 | .27 | .39 | .34 | .48 | .63 |
| Plate glass | .18 | .06 | .04 | .03 | .02 | .02 |
| Plaster on lath on studs | .30 | .15 | .10 | .05 | .04 | .05 |
| Acoustical plaster (1 in) | .25 | .45 | .78 | .92 | .89 | .87 |
| Plywood on studs (¼ in) | .60 | .30 | .10 | .09 | .09 | .09 |
| Perforated cane fiber tile, cemented to concrete, ½-in thick | .14 | .20 | .76 | .79 | .58 | .37 |
| Perforated cane fiber tile, cemented to concrete, 1-in thick | .22 | .47 | .70 | .77 | .70 | .48 |
| Perforated cane fiber tile, 1-in thick, in metal frame supports | .48 | .67 | .61 | .68 | .75 | .50 |

Furthermore, that sound was very "dead," a matter to which we shall return.

Besides increasing the intensity of sounds, reflections from walls help us judge the distance of a sound source. For the same intensity of sound reaching an observer, the distance of the source is judged to be greater if the same sound comes from several loudspeakers at different distances rather than from a single speaker. This is quite reasonable. If someone speaks to us in a room, we hear mostly direct sound if the person is close to us, mostly reflected or reverberant sound if he or she is far away. In an anechoic chamber (a room whose walls do not reflect sound), if you stand behind a person and whisper, the person will think that you are close, regardless of distance. John Chowning has used varying amounts of reverberation to give a sense of varying distance to computer-generated sounds in his compositions *Stria* and *Turenus*.

Playing in a "dead" room can be exasperating to the performer. Musicians who played in Philharmonic Hall (now Avery Fisher Hall) in Lincoln Center before

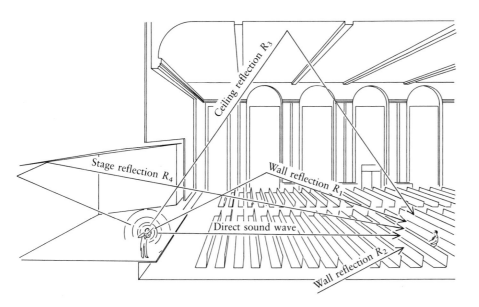

extensive modifications were made found it difficult to hear themselves play, because the wall behind the performers was made of sound-absorbing material (see the table on the facing page). I've been told that performers like to play about 15 or 20 feet in front of a reflecting surface. They can hear themselves play. This seems natural to them and helps them.

Reflected sound, or reverberation, is even more important to audiences. I once heard a recording of a pipe organ that had been made in the organ loft. It sounded like an electronic organ, because there were no reverberations. Reverberation blends and adds richness to sounds. My wife and I once heard a student orchestra practicing in Sainte Chapelle in Paris. They produced a shimmering cascade of sound, especially from the trumpets. "It sounds wonderful," I said. "But they aren't together," my wife replied. Indeed, they weren't, but the individual notes were so lost in grand sonorities of reverberation that I hadn't noticed.

Reverberation time is defined as the time it takes a sound to decrease to 60 dB below its initial intensity. Speech becomes hard to understand when the reverberation time is greater than one second, and is clearer when the reverberation time is half a second. Two seconds is fine for music, and some modern organ music seems designed for cathedrals with still longer reverberation times (see the figure on page 141, Chapter 11).

What about echoes? Echoes are very annoying. As we have noted, when we hear the same sound from two sources 20 to 26 meters (60 to 80 feet) away (a time difference of 60 to 80 milliseconds), we hear an echo. We hear an echo in an auditorium if we get one strong, delayed reflection from a flat surface, such as the front of a balcony. But in a well-designed auditorium, single strong reflec-

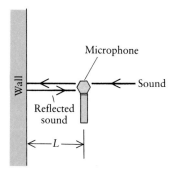

A microphone in front of a perfectly reflecting wall. Sound reaches the microphone with equal intensity, both directly and after being reflected from the wall.

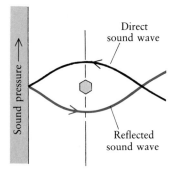

Direct and reflected sound waves for the microphone in the figure above. If the frequency of the sound is $v/4L$, as shown, the pressure of the reflected sound wave will be equal and opposite to the pressure of the direct wave; so the microphone will pick up nothing.

tions are prevented. Instead, we get a multitude of reflections that reach our ears at different times and from different directions. Because the sound reaches our ears first by a direct path without reflections, we judge the whole of the sound to come from a small source on stage. However, its quality is very different from what it would be without reverberation.

Strong, single, distant reflections are bad. So are small nearby reflections. We are told to put loudspeakers either against or in walls, or else far away from them. The effects of close reflections are even more apparent with microphones. If you put a microphone two or three inches from the surface of a table, whatever it picks up, voice or music, sounds bad. Sounds are "colored." We hear some frequencies as emphasized, some as suppressed.

We can easily see why this is so. The first figure at the left shows a microphone in front of a perfectly reflecting wall. A sound coming from the right reaches it twice, once directly and once reflected, at a time

$$t = \frac{2L}{v} \text{ seconds}$$

later, where $v$ is the velocity of sound.

Consider the second figure at the left. If the sound is a tone of frequency

$$f = \frac{v}{4L},$$

a trough of the reflected wave (in color) will reach the microphone just as a crest of the original wave reaches the microphone. The two waves, direct and reflected, will cancel; so the microphone won't pick up anything at that frequency. The response of the microphone to tones of increasing frequency will also go to zero at frequencies $v/4L$, $3v/4L$, $5v/4L$, and so forth, as shown at the top of the facing page.

We can prevent the bad effects of reflected sound by putting the microphone very close to the wall. We can also prevent gross coloration of sound by putting the microphone far from the wall, though the sound will be affected by the reverberation of the room.

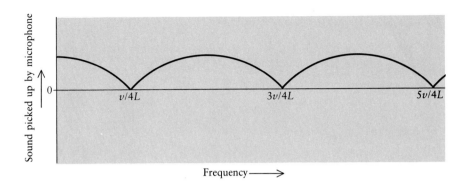

Sound picked up by microphone

0 — $v/4L$ — $3v/4L$ — $5v/4L$

Frequency ⟶

 How the sound picked up by the microphone of the lower figure on the facing page varies with the frequency of the sound wave.

We tend to hear reverberation rather than coloration when the difference between frequencies at which the response goes to zero or dips sharply is small compared with a critical bandwidth (roughly a minor third, or 5/4 of the frequency in question). The frequency difference between successive dips in the response curve in the figure above is $v/2L$. Hence, for reflections not to distort the sound, we should have

$$\frac{v}{2L} \text{ smaller than } (5/4)f$$

or

$$L \text{ larger than } (2/5)(v/f).$$

If we want to prevent distortions for frequencies down to 100 Hz, we must make $L$ at least three meters. (Actually, the critical bandwidth is greater at lower frequencies; so three meters is larger than needed.) Anyway, you can now see why you must keep your microphone well away from floors and walls when recording music.

One other matter concerning the perception of music is worth mentioning: our ability to distinguish small periods of time. In 1973 David M. Green published some interesting results on temporal acuity. He measured the ear's ability to discriminate between two signals that have the same energy spectrum. An example of such signals is any short waveform and the same waveform reversed in time, such as those in the adjacent figure. Here part A is a sound of decreasing frequency, part B of increasing frequency. Green found that the ear can tell the difference between two such waveforms if their duration is greater than 2 milliseconds.

The phenomena of the precedence effect and of echoes are of extreme importance to music. If one of two sources of the same sound lags behind the other by more than a millisecond (or foot), the earlier source "swallows up" the later as far as our sense of direction goes. But if the time difference is more than 60 to 80 milliseconds, we hear an echo. Although we sense a sound as coming from the direction of first arrival, later arrivals add both intensity and a reverberant quality, which is essential for good musical sound.

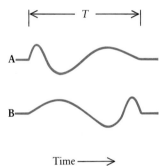

⟵ $T$ ⟶

A

B

Time ⟶

■ Two sound waves (A and B) have different waveforms but the same energy spectrum. Measurements made by David M. Green show that such waves are heard as different if their duration is greater than 2 milliseconds. Here wave B is wave A reversed in time.

# Architectural Acoustics

Between 1895 and 1915, Wallace Clement Sabine, Hollis Professor of Mathematics and Philosophy at Harvard University, laid the foundations of a new science, architectural acoustics. Before Sabine, good acoustical design consisted chiefly of imitating halls in which music sounded good. Poor acoustic design consisted of superstitious practices, such as stringing useless wires across the upper spaces of a church or auditorium.

Architectural acoustics was founded because an opportunity was presented to a remarkable man. The opportunity arose because it was almost impossible to understand speakers in the lecture room of the newly opened Fogg Art Museum. In 1895, the Corporation of Harvard University asked Sabine to remedy this.*

Sabine approached the problems of architectural acoustics with a sharp and inquiring mind, a keen ear, a stop watch, and an organ pipe with a tank of compressed air as a source of sound. He identified the persistence of sound (i.e., the excessive reverberation) in the Fogg lecture room as the factor that rendered speech unintelligible. He reduced this reverberation by placing felt on particular walls. This, he said, made the room "not excellent, but entirely serviceable."

Sabine was the first to define *reverberation time,* one important parameter of lecture halls and auditoriums. His definition was the time that it takes, after a sound is turned off, for the reverberant sound level to become barely audible. When accurate electronic measurement of sound level became possible many years later, this turned out to be a fall in sound level of 60 dB, which is how reverberation time is defined today. What reverberation time is optimum depends both on the type of music or other sound to be heard and on the size of the room, theatre, or other enclosure. The figure on the facing page shows proposed optimum reverberation times for various purposes plotted against room volume.

Accurate calculation of reverberation time has been a persistent problem of architectural acoustics. Sabine not only first defined reverberation time, but also devised a useful if not perfectly accurate way of computing it in terms of volume and the fraction of the incident sound that walls and other surfaces reflect.

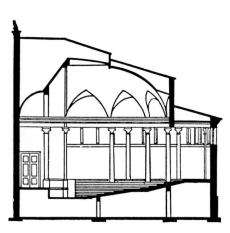

■ Fogg Art Museum lecture hall.

■ The photograph on the preceding pages is of Louise M. Davies Symphony Hall in San Francisco, a tunable hall.

---

*The story of what he did can be read in his own words, if you can find a copy of his *Collected Papers on Acoustics,* first published by Harvard University Press in 1922, and reprinted by Dover in 1964, but now, unfortunately, out of print.

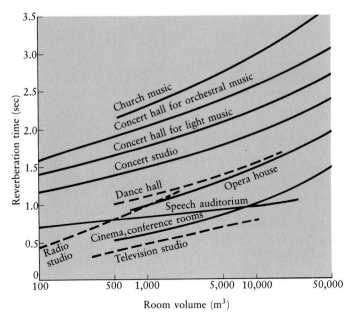

The best reverberation time is different for different uses, and for a given use, increases with the volume of the enclosure. Church music sounds good in huge, highly reverberant cathedrals. If a room were cubical, the 100-cubic-meter room would be 4.6 meters or 15 feet on a side, the 50,000-cubic-meter room would be 38 meters or 125 feet on a side.

Sabine's work was arduous. Some was spent in an underground room with brick and concrete walls, where, seated in an enclosure, he measured the sound-absorptive properties of various materials and made other studies. Elsewhere, he spent long nights, waiting for periods quiet enough that he could get an absolute calibration of the sound-absorptive properties of materials by comparing their effect with that of open windows. Sabine measured the sound absorption of a host of materials. He also studied sound conduction and means for isolating practice rooms acoustically, a persistent problem in conservatories.

Mostly, Sabine was called on to cure or ameliorate the bad acoustics of halls built by presumably respectable architects who were either grossly ignorant of the needs of people or who simply didn't care about them, traits still all too common among architects today. But Sabine was also able to do an original acoustical design of Symphony Hall in Boston, built in 1900 to replace the old Music Hall. Today, Symphony Hall is one of the very few outstanding concert halls in the world.

Research in architectural acoustics has waned in the land of its birth, but excellent work continues in West Germany and other countries. There are two general problems in architectural acoustics, and each has many aspects. One problem is, what do we want? What enables performers to play well? When they do play well, what is it that makes them sound good? The other problem is, how can we attain what is good for the performers and what is good for the audience?

Noise is important in concert halls, as noted in Chapter 9. Concert-hall design requires great attention both to excluding external noise and to not producing noise. For example, air-conditioning systems are often excessively noisy in offices, and sometimes are in concert halls.

The performer needs to hear reflected sound, as noted in Chapter 10. This has been the subject of several recent studies, both of performers' preferences for various halls and of performance under laboratory conditions. Although any reverberant sound will satisfy soloists, satisfactory ensemble playing depends on early reflections of sound from behind and above the performers. Each player must hear all the rest by means of reflected sound that is not too much delayed. Good transmission of sound from the orchestra to the audience will be of little avail if the players can't play well and comfortably together. It is common lore that players and even audiences like to "feel" music through a wooden (as opposed to a concrete) floor.

The problem of out-of-doors performances, such as those in the Hollywood Bowl, are as old as the Greek theater. Such performances are plagued by noise. Although it is easy to provide reflecting surfaces near the orchestra to enable the performers to hear one another, it is impossible to provide reverberant sound to the audience, at least, not without amplification and artificial reverberation. Without electronic aid, outdoor music *can't* sound as good as music in a concert hall. It may be fun, but the reasons it is fun aren't good acoustics.

In this chapter we will consider in some detail both the physical aspect of measuring and predicting the transmission and decay of sound in concert halls, and the psychological aspect: what makes a good hall good? Sabine appreciated both of these aspects. He experimented to find the preferred reverberation time for musical performance, and he knew that the optimum reverberation time for music is longer than that for speech. Sabine also understood the objectionable quality of echoes, and knew how to prevent or cure them.

■ Philharmonic Hall, as designed by Leo Beranek.

Much progress has been made since Sabine's time. Some has hinged on new psychoacoustic understanding (e.g., of the precedence, or Haas, effect). Electronic means for generating and measuring sound have been most useful. The electronic digital computer has been of great help. But these resources would have meant nothing without the sharp, inquiring minds of a few men who have, through the years, advanced the science of architectural acoustics.

How far has it advanced? In a practical sense, not as far as we might like. Philharmonic Hall in Lincoln Center, Manhattan, opened on September 12, 1962. It was a disaster. Yet, in July of 1962, Leo L. Beranek, of the architectural firm of Bolt, Beranek, and Newman, wrote in the preface of his book *Music, Acoustics, and Architectural Design,* "The climax of this volume is the description of the care taken in planning the Philharmonic Hall in Lincoln Center. Lady Luck has finally been supplanted by careful analysis and the painstaking application of new but firmly grounded acoustic principles."

What went wrong in the original design of Philharmonic Hall, and why?

The *why* appears to be that great attention was given to matters that Beranek deemed to be of great importance, whereas little was given to other matters that also proved to be of great importance.*

---

*Beranek was so convinced of the correctness of his theories that he did not even bother to build and test a model of the hall. Models have been made since Sabine's day, and his book shows pictures of sound propagating through a model auditorium. What Sabine could do crudely can now be done well. The wavelength of sound in a model should preserve the same relation to dimensions as in the actual hall; for example, in testing a tenth-scale model, a frequency of 500 Hz should be represented by a frequency of 5,000 Hz.

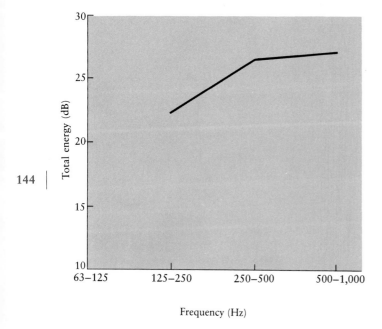

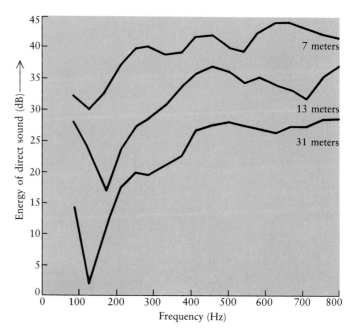

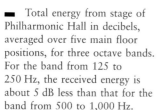 Total energy from stage of Philharmonic Hall in decibels, averaged over five main floor positions, for three octave bands. For the band from 125 to 250 Hz, the received energy is about 5 dB less than that for the band from 500 to 1,000 Hz.

■ Energy of sound going directly from stage of Philharmonic Hall to various locations along the center aisle on the main floor, as a function of frequency. At 31 meters from the stage, the sound at 130 Hz is about 25 dB below that at 500 Hz.

The *what* includes almost everything. (The details can be found in papers by Manfred Schroeder and his colleagues.) There were echoes at some seat locations. The members of the orchestra couldn't hear themselves and others play, because the wall behind them was absorptive. There was a lack of subjectively felt reverberation. There was inadequate diffusion of sound through the hall. Worst of all, there was an apparent absence of low frequencies: it was difficult to hear the celli and double basses.

This was not apparent in measurements of the total sound energy that reached the hearer during the period of one second after the first arrival of direct sound. The figure at the left above shows this total energy in various octave bands of frequency. Though the energy has fallen somewhat in the octave from 125 to 250 Hz, the drop is only about 5 dB, which isn't much.

What *was* wrong was that there was indeed an initial lack of low-frequency sound, both in the sound that arrived directly from the stage and in sound reflected from a suspended ceiling made up of a large number of adjustable panels or "clouds."

The figure at the right above shows the deficiency in the direct sound from the

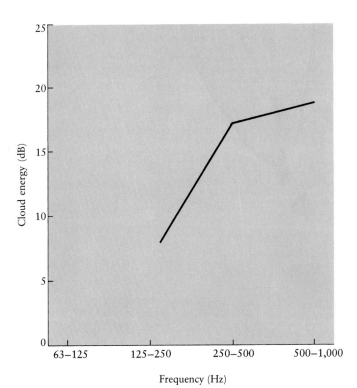

■ Energy in octave bands reflected from clouds, like those in Philharmonic Hall, averaged over five main floor positions. This measurement was made in a model. The energy reflected in the band from 125 to 250 Hz is about 10 dB below that reflected in the band from 500 to 1,000 Hz.

stage. The relative energy of the direct sound in dB is plotted against frequency for distances of 7, 13, and 31 meters from the stage, along the center aisle of the main floor. There is a pronounced dip between 100 and 200 Hz, and it becomes greater at greater distances from the stage. The original floor of the hall wasn't raked much; that is, it didn't much go downhill toward the stage. The sound skimmed along above the rows of seats. Because the space between rows of seats tended to act as a resonator, at low frequencies the sound waves were bent upward, away from the audience. The way the sound waves were reflected from the regular, successive rows of seat backs (diffraction) added to the loss of low frequencies.

What of sound reflected from the ceiling? The figure above shows the average energy reflected from the "clouds" during a short interval centered on the arrival time from the clouds. For the octave from 125 to 250 Hz, the average energy has fallen about 11 dB below its value at the octave from 500 to 1,000 Hz.

Why was so little low-frequency energy reflected from the clouds? Because they weren't large enough! A flat surface acts as an effective reflector only if it is large measured in wavelengths of sound. The clouds weren't large enough to effectively reflect sounds having frequencies below 300 Hz.

The curve of the right-hand figure on the facing page shows how little direct sound of low frequency reached the main floor. The figure above shows how little was reflected from the clouds. Why, then, does the left-hand figure on page

144 show a considerable amount of low-frequency sound energy finally reaching the listener? It reached the listener only after repeated reflections, some above the clouds, which low-frequency sounds passed through, and some after traversing the hall many times. This sound arrived so late that the listener failed to associate it with the notes that the celli and basses were playing. It became, in effect, a background noise, detached from its musical source.

There were many early efforts to patch up Philharmonic Hall. A solid stage enclosure was built so that the orchestra could hear itself play. The clouds were realigned to form an essentially continuous ceiling. Scattering elements were put on the side walls to give better sound diffusion. New, less absorptive seats were installed on the main floor. The front of the balcony was tilted, and absorptive material was placed on the back wall to diminish echoes. This made the hall somewhat better, though the reverberation time became rather low (about 1.85 seconds).

Finally, Philharmonic Hall was completely redesigned by one of the few first-rate American experts in architectural acoustics, Cyril Harris of Columbia University, and became Avery Fisher Hall.

The story of Wallace Sabine is that of a triumph of new scientific understanding. The story of Philharmonic Hall is that of an expensive disaster based on incomplete knowledge. There is a happier side to architectural acoustics, well-illustrated by the work of Manfred Schroeder, whose contributions illustrate what progress in architectural acoustics can be.

A persistent problem in designing concert halls has been the accurate prediction of reverberation time. Sabine gave a simple formula for the reverberation time $T$, measured in seconds, as

$$T = \frac{13.8L}{va}.$$

Here $L$ is the mean free path between successive reflections of sound waves, $v$ is the velocity of sound, and $a$ is the sound-absorption coefficient, which is zero for perfect reflection and unity for complete absorption.

Sabine assumed that the mean free path $L$ was proportional to the cube root of the volume. However, it was already known from the kinetic theory of gases

■ Cyril Harris.

■ Avery Fisher Hall on opening night.

that under what is called an *ergodic* condition, in which the sound traverses all
possible paths, the mean free path is given by

$$L = \frac{4V}{S}.$$

Here $V$ is volume, and $S$ is the internal surface area of the volume.

A problem with Sabine's formula is that it predicts a finite reverberation time
for complete absorption ($a = 1$). In 1929, K. Schuster and E. Waetzmann, and
in 1930, Carl F. Eyring remedied this by proposing a revised formula of

$$T = -13.8(L/v) \ln (1 - a)$$

The Sabine/Eyring formulas were accepted for more than half a century, but, in
the 1960s, accurate electronic measurements of various halls, new and old,
cast doubt on them.

What to do? One approach was to use the digital computer to trace many sound
paths or rays through successive reflections in an enclosure. Some rays would
encounter absorptive material and be reflected with loss of energy; others would
be reflected from wood or plaster with little loss in energy. In 1970 Schroeder
published a paper giving examples of reverberation times derived by ray tracing
in irregular two-dimensional enclosures with absorbing material on various
walls. He compared the slope of the curve of sound level in dB against time
(which gives the reverberation time) with slopes calculated according to the
Sabine and Eyring formulas. The figure at the left on the next page shows the
comparison. For the trapezoidal enclosure shown, with sound-absorbing mate-

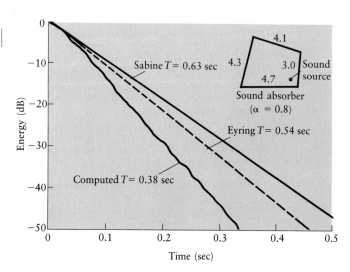

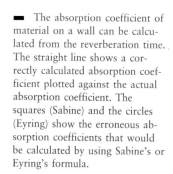

Decay of sound computed by ray tracing for a trapezoidal enclosure with absorbing material on one wall. This computed curve corresponds to a reverberation time of 0.38 seconds. The reverberation time calculated according to Sabine is 0.63 seconds, and according to Eyring is 0.56 seconds. The Sabine and Eyring calculations do not take into account the fact that all sound absorption is on one wall.

The absorption coefficient of material on a wall can be calculated from the reverberation time. The straight line shows a correctly calculated absorption coefficient plotted against the actual absorption coefficient. The squares (Sabine) and the circles (Eyring) show the erroneous absorption coefficients that would be calculated by using Sabine's or Eyring's formula.

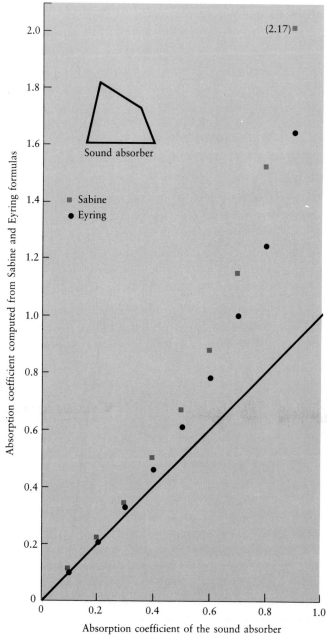

rial on one wall, the computer ray-tracing calculation gives a reverberation time that is 0.6 of that predicted by Sabine's formula, and about 0.7 of that predicted by Eyring's.

Such results are important not only for the design of concert halls, but also for the measurement of the absorption coefficient of sound-absorbing materials. The figure at the right on the facing page illustrates this.

For a given shape of enclosure, the rate at which sound dies down depends on the dimensions of the enclosure and on the amount and position of the absorptive material. If we measure how sound dies out and compare this with a correct calculation of reverberation time, we thus measure the absorption coefficient of the absorptive material used. In the right-hand figure on the facing page the coefficients of absorption computed from Sabine's and Eyring's formulas are compared with that computed by ray tracing. If we accept the results of ray tracing, we must believe that Sabine's or Eyring's formula leads to serious overestimates of the sound absorption of acoustic materials. Indeed, measurements could lead us to more than 100% sound absorption.

More recently, further progress has been made in the theory of reverberation time. For many years it has been known that a nasty integral equation provides the "proper" approach to reverberation. The equation is essentially unsolvable, even by using a computer. Recently, E. N. Gilbert of Bell Laboratories showed that, by using another valid integral equation as well, he could obtain a solution by an iterative process. Gilbert has thus obtained "correct" results for several enclosures having simple but interesting shapes.

Calculations or measurements of physical quantities such as reverberation time are all very well, but what we want to know is, how good does an orchestra sound in a particular hall? Judgment is difficult, for different orchestras play in different halls at different times, and the mood of the peripatetic listener may differ on different occasions. Is there any way (outrageous thought!) to bring different halls to one listener at the same time? There is!

In 1967, Manfred Schroeder and Bishnu Atal showed how two loudspeakers could be used to produce an apparent sound source that lay to the left or right of both speakers. In 1969, P. Damaske and V. Mellert showed how to make use of this effect in producing a "perfect" stereo system. Sounds were picked up

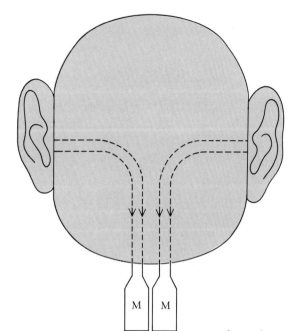

A dummy head with pinnae. Microphones (M) pick up the sound pressure in the ear canals. Two-track stereo recordings can be made in a concert hall by "seating" the dummy head in the hall.

If the two-track recording made with the dummy head shown above is played back by means of two loudspeakers in an anechoic room, with the use of a proper network so that some signal from each sound track reaches each loudspeaker, the pressures in the ear canals of a real (or dummy) head will be just those recorded in the ear canals of the dummy head seated in the concert hall. The fraction of the signal fed from the microphone in the left ear of the dummy head to the left and right loudspeakers must change properly with frequency.

from the two ear canals of a dummy head, complete with pinnae. The figure at the left shows such a dummy head. These signals were then filtered and mixed properly, and fed to two loudspeakers in an anechoic room. The diagram on the facing page indicates how the sound was reproduced from the two-track recording. The sounds from the two speakers could recreate in the ear canals of a dummy head exactly the same sound pressures that had been recorded when the head was exposed to live sound during the recording.

Of what use was this? Suppose that one made a two-channel recording of the sounds in the ear canals of a dummy head "seated" in a concert hall. From the recording one could recreate for a listener in an anechoic room *exactly what the dummy head heard in the concert hall.* By switching from a recording in one hall to a recording in another hall, one could compare halls.

Schroeder and two collaborators, D. Gottlob and K. F. Siebrasse, undertook to compare more than twenty European concert halls. They managed to obtain a multichannel tape recording of Mozart's *Jupiter* symphony played by the BBC orchestra in an anechoic room. This tape they played back over several loudspeakers on the stages of various concert halls. In each hall they made two-channel tapes of what the dummy head heard when seated in several locations. From these recordings they recreated for a listener seated in an anechoic room exactly what the dummy head heard in various concert halls. Schroeder comments,

> I will never forget the moment when, comfortably seated in the Goettingen "free space" room, I first switched myself from Vienna's famed Musikvereinsaal to Berlin's Philharmonie. The acoustical differences of these halls, always believed to "exist," stood out in a manner so vivid that it is difficult to put into words.

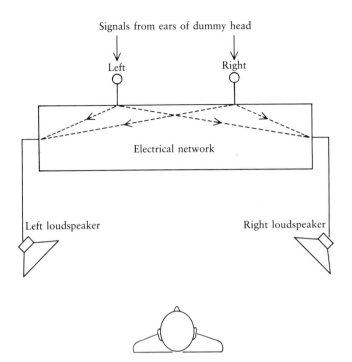

Signals from ears of dummy head

Left      Right

Electrical network

Left loudspeaker      Right loudspeaker

■ Dummy head and D. Gottlob (left) and K. F. Siebrasse (right).

By analyzing the judgments that various listeners made "listening" in many positions in many halls, Schroeder and his collaborators learned what listeners liked:

1. They liked long reverberation times (below 2.2 seconds).
2. They liked the sounds to differ at their two ears. The more nearly alike (correlated) the sounds were at the two ears, the less they liked it.
3. They liked narrow halls better than wide halls. Perhaps this is another expression of a preference for different sounds at the two ears. In a wide hall the first reflected sound rays reach the listener from the ceiling. In narrow halls the first reflections reach the listener from the left and right walls, and these two reflections are different.

What, then, must a good hall do? It must have a long enough reverberation time. Beyond that, it must mix the sound up, so that the sounds reaching the two ears are different. Such mixing up can also help to make all seats good, for it favors diffusion of sound through the hall.

■ Hall designed by Manfred Schroeder at IRCAM, Paris.

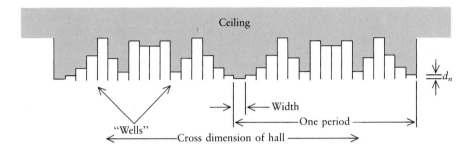

■ A sequence of "wells" that scatters incident sound widely without loss, rather than reflecting it like a mirror. The scattering changes somewhat with frequency but is good over a wide band of frequencies. Material with a similar surface may be used in concert halls of the future to insure diffusion of sound throughout the hall and to distribute the sound so that the sounds reaching the two ears are different.

Much mathematics has been expended recently in designing textured walls or ceilings with bumps and wells that will scatter incident sound in many directions rather than reflecting it as a mirror does. The figure above shows an example in which the depths of long, narrow wells in a ceiling are based on "quadratic residues" derived from number theory. If the ceilings and walls of some new concert hall are strangely rough and pockmarked, don't be surprised.

Don't be surprised, either, by *tunable* halls, in which the reverberation time can be adjusted, either by shifting absorptive panels or by electronic means, to suit different kinds of music. The new Louise M. Davies Symphony Hall of the San Francisco Symphony is mechanically tunable, and so is the *Espace de Projection* at IRCAM in Paris.

When does music sound good? In a good hall. What are the characteristics of a good hall? It must be quiet. It must be reasonably good for all seats. Sounds of all frequencies must reach the listeners. The amount of reverberation must suit the use; more is better for some sorts of music than for others. Theaters intended for plays should be less reverberant. For music to sound good, the hall must mix up sounds, so that what reaches the left ear is different from what reaches the right ear. Perhaps there is more than that. Those who study architectural acoustics continue to experiment and learn.

# Sound Reproduction

The electronic reproduction of music has become far more common than live music. Pop is the best seller, but the recordings of "good" music are extensive. Once, such music could be heard only in the excitement (or boredom) of a concert hall. Today the concert hall is a rare and expensive experience. Live performances can't keep up with an increasing market. The sizes of halls and their audiences can grow somewhat, but not without limit. It is hard, and some say impossible, to design a very large hall that has good acoustics. Some claim that a 1,500-seat hall is optimum. Construction costs and the salaries of musicians and other personnel have made halls of that size completely uneconomical. Indeed, whatever the size of the hall, a symphony orchestra that does not make money from recordings, TV, or films must survive on charity or government subsidies; yet people hear more music, and a greater variety of music, than ever before.

How well do they hear it? Some don't care much. A tinny transistor radio can remind one of a symphony orchestra and the composition it plays. I can enjoy conventional works that I have heard before even when they are poorly reproduced, though nonlinear distortion bothers me.

It is harder to judge or enjoy unfamiliar music or performances when the music is poorly reproduced. It is particularly hard to appreciate computer-generated music, which may depend on deep bass tones or on stereo effects that may take the music on a ghostly course around the room. Such sounds are designed to be heard on a good stereo system, which will reproduce them well.

What about large orchestras? If we spend enough money, will we hear them as loudly, as clearly, as well as we wish? That depends on what we ask of sound and where we hear it. On April 27, 1933, Harvey Fletcher and his co-workers demonstrated a three-channel system that carried the sound of the Philadelphia Orchestra, conducted by Alexander Smallens in the Academy of Music in Philadelphia, to three loudspeakers on the stage of Constitution Hall in Washington, D.C. During the demonstration Dr. Leopold Stokowsky manipulated the electronic controls from a director's box in the rear of Constitution Hall. The figure on the facing page shows the placement of the microphones in the Academy of Music and of the loudspeakers and the director's box in Constitution Hall.

The performance of this system was outstanding by any standard. The bandwidth was from 40 to 15,000 Hz. The range between noise in the absence of

The photograph on the preceding pages is of Rosario Bourdon conducting the Victor Salon Orchestra in an acoustic recording session, ca. 1920.

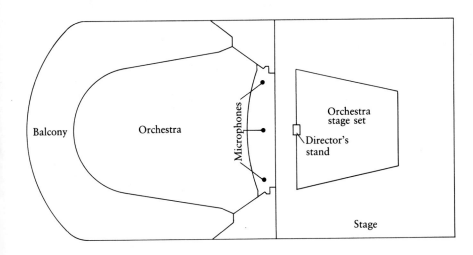

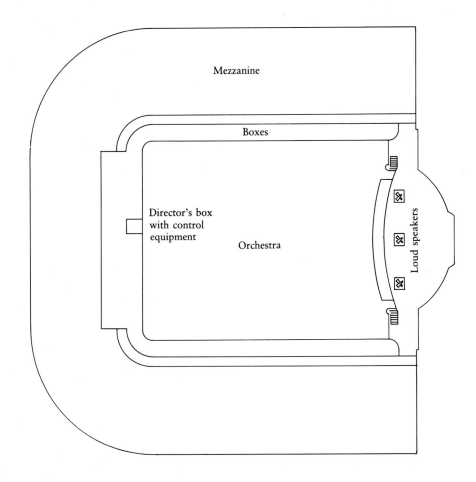

■ On April 27, 1933, Fletcher and his co-workers transmitted the sound of a symphony orchestra playing in the Academy of Music in Philadelphia (top) to Constitution Hall in Washington, D. C. (bottom). The sound was picked up by three microphones between orchestra and audience, and reproduced by means of three loudspeakers on the stage of Constitution Hall. In a director's box near the back of Constitution Hall, Leopold Stokowsky adjusted sound levels during the performance.

■ Constitution Hall, Washington, D.C.

signal and the highest average signal power was 75 dB (30,000,000 times). The largest undistorted sound-output power that each of the three loudspeakers could deliver was 135 watts, or 405 watts for all three loudspeakers. Each loudspeaker consisted of a folded horn woofer with a square aperture about five feet on a side to handle frequencies from 40 to about 300 Hertz. This was surmounted by a cellular horn about two feet high and three feet wide to handle higher frequencies.

The acoustic power that this system could deliver was several times that which a large orchestra can produce, which is about 70 watts. I've heard that Stokowsky kept turning the sound up louder and louder, and that the controls had to be modified to keep him from overloading the system and blasting the audience. Whatever the truth of this may be, by all accounts the effect was good. The sense of the sources of the sounds of individual instruments cannot have been preserved, but in most seats in a large hall we are so far from the stage that we scarcely sense more than left, center, or right. Mostly we are enveloped in a field of reverberant sound from the walls and ceiling, and it seems that little was lost because the sound came from three loudspeakers placed appropriately on the stage.

In experiments described in Chapter 11, Schroeder obtained a very convincing sense of the presence of an orchestra by playing a multichannel recording through loudspeakers placed on the stages of concert halls. Schroeder's reproduced sound should have been better than Fletcher's, for the recording was made in an anechoic room, whereas in Fletcher's system some reverberation from the first hall must have got into the sound sent to the second.

If we want the effect of an orchestra in a large hall, we can get a pretty good approximation, but there can scarcely be a large market for such a system, not unless Walt Disney's 1940 *Fantasia* is imitated and repeated many times. *Fantasia* coupled Disney's animation, some of it abstract, with Stokowsky and the Philadelphia Orchestra. In the original showing, loudspeakers at the sides of the theatre supplemented those on the stage in producing special effects.

When we think of sound reproduction, we usually think of reproducing the sound of a singer, instrumentalist, ensemble, or orchestra in our own home. Can this be done? To a degree. In Chapter 11, I described a perfect two-track stereo system, in which sound pressure in the ear canals of a dummy head is faithfully reproduced in the ears of a listener seated in an anechoic room. We should note three limitations.

First, this system can work perfectly *only* in an anechoic room. In any other enclosure, the reverberation of the enclosure is added to that of the hall in which the sound was recorded. This may not be very important, because the precedence effect insures that you will at least hear the sound coming from the right direction.

Second, you can't turn your head very much. Third, you must be equidistant from the two loudspeakers, and at just the right distance from them. If you move close enough to one loudspeaker, any sound that comes from both speakers will seem to come from the loudspeaker to which you are nearest. If you move a little, you may hear sound from both speakers, but phantom auditory images of compact sound sources will move and become blurred.

Despite these limitations of the "perfect" two-channel system, something may be gained by a crude use of one of its features, that of driving each speaker from a mixture of the two recorded sound channels, as shown in the diagram on page 151. Hi-fi companies supply networks that do something of the sort, and the effect seems to be an increase in "presence," or being surrounded by sound. Of course, this cannot approximate the "perfect stereo" effect, for the two sound channels that are mixed were not recorded with the ears of a dummy head.

159

■  The type of speakers used in the 1933 transmission from Philadelphia to Washington.

Ludwig II, king of Bavaria, had an opera house with one seat for an audience of one, himself. Many of us would prefer to share performances, live or recorded, with others. Is there any way to share electronically reproduced sound "equally"? With headphones, perhaps; we will come to that later. Using loud-speakers, it is possible in principle, but not in practice, for the following reasons.

Imagine a room-sized enclosure built in a concert hall, with musicians playing on the stage outside the enclosure, and yourself seated inside the enclosure.

If all walls, the floor, and the ceiling are solid and soundproof, without aper-tures, you will hear nothing.

If we cut one hole in a wall, you will hear all the music as coming from that hole, as you would from a single loudspeaker.

If we cut two holes in the wall, you will get an effect much like that of a conventional stereo system. The music will sound like good stereo if you sit equidistant from both holes, but you will never hear anything coming from a direction to the right or the left of the two holes (loudspeakers), as you can with the "perfect" stereo system in which sound is picked up by a dummy head, and you will have no sense of sound coming from above or below the holes.

If we cut a hole in every wall, you can sense sound as coming from any direction, but again, if you move close to one hole, you will hear sound as coming from that hole, and unless you are equidistant from the four holes, you will not hear sounds as coming from their original compact sources in the correct directions.

Only if we cut many holes in the walls and ceiling can we hear sound within the enclosure pretty much as we would hear it in the absence of the enclosure. We could approximate this effect by putting many, many loudspeakers on the walls and ceiling of a room, and feeding an amplified signal to each loudspeaker from a microphone placed in an analogous position in a concert hall. Such a multi-channel system would be costly, unwieldy, and impractical.

What can we hope to do in practice? What we can hope to do really well is to reproduce the concert-hall sound at one position in a room, marred only by reverberation in the room. We can do this with the two-channel system using a dummy head. That isn't what hi-fi enthusiasts do if they aren't satisfied with conventional stereo. Rather, they use a four-channel or quadraphonic system, though the advantages of quadraphonic sound over good conventional stereo are questionable. Indeed, some in the hi-fi community denounce quad systems in strong terms.

A quadraphonic system does not attempt to accurately reproduce the sound pressures at the two ears of the listener. Rather, at its most accurate it can reproduce at one point in a room the exact sound pressure and velocity that exist at some point in the recording studio.

Recall from Chapter 2 that a sound wave consists both of motion of the air and of compression and rarifaction of the air. At any point, a sound wave, no matter how complicated, disturbs the air in just four ways: by a fluctuating pressure, by a fluctuating up-and-down movement or velocity, by a fluctuating forward-and-back velocity, and by a fluctuating left-and-right velocity. Some microphones are sensitive to pressure, some to velocity in a particular direction. In principle, at least, we can measure the fluctuations of pressure and of the three velocities of *any* sound wave at one point in the recording studio. Over four audio channels or via a four-track recording, we can transmit electric signals representing the pressure and the three velocities to a room in our own home.

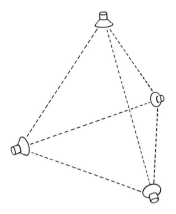

■ The exact sound velocities and pressure at a point in a recording studio can be reproduced at a point in an anechoic room by means of four loudspeakers placed at the corners of a tetrahedron. Four microphones in the studio pick up signals proportional to (1) the up-and-down velocity of the air, (2) the forward-and-back velocity, (3) the left-and-right velocity, and (4) the pressure. The signal that drives each loudspeaker is a particular linear combination of these four signals.

In that room we can place four loudspeakers. Plausible positions for these loud-speakers are the four corners of a tetahedron, a four-faced pyramid with equi-lateral triangles as sides and base (as shown at the right). Let us imagine the listener to be at the center of the tetrahedron. A particular loudspeaker will produce a sound wave that has a fluctuating pressure and a fluctuating velocity toward and away from that loudspeaker. By feeding the proper signals to all four loudspeakers, we can produce any desired pressure fluctuation at the center of the tetrahedron, and any desired velocity fluctuation in any direction.

Indeed, by driving each loudspeaker with a signal that is a linear combination of the four signals that represent the pressure and the three velocities at a point in the recording studio, we can reproduce *at the center of the tetrahedron* the same fluctuations of pressure and of the three velocities that were measured at one point in the recording studio. At one point between the loudspeakers, the fluctu-ations in air pressure and velocity will be exactly the same as those at the micro-phones in the recording studio. The argument is that, if you place your head at this point (and if your head isn't too large), you will hear exactly what you would hear in the studio.

The arrangement of loudspeakers at the corners of a tetrahedron can give a sense of the height as well as of the horizontal direction of the source of a sound wave. Actual quadraphonic systems give no such sense, because they use four loudspeakers at the corners of a square. Although four channels are used in true quadraphonic recordings, only three channels are really needed for such a system, for it can reproduce only the pressure, the forward-and-back velocity, and the left-and-right velocity at some point in the studio. In matrix quad, four channels are combined and recorded as two independent channels, and four channels are then derived from these two channels. Matrix quad works (sort of) because the signals recorded from four microphones arranged in a square aren't completely independent.

Quadraphonic systems use more transmission or recording channels than the two necessary for the "perfect" stereo system using a dummy head. Why, then, are quadraphonic systems used? Largely, I think, because they were thought of first, and partly because the stereo industry hasn't been closely in touch with the work of psychoacousticians. There is another reason.

Four channels (or even two) allow us to achieve a variety of effects. If the sound of one voice or one instrument comes from one loudspeaker only, it will come from the direction of that loudspeaker, no matter where we stand in the room. Multichannel recording or transmission and reproduction *can* be used to give all those in the room a sense of being surrounded by sound. According to Schroeder's studies (see Chapter 11) that is what people like in a concert hall.

What do people want of sound reproduction? What do they like? Some hi-fi purists actually want to reproduce at the listener's ears just what one would hear in a concert hall. We have noted that this is impossible except at one point in an anechoic room, though we can come close to it in an ordinary room. However, that isn't what people who listen to popular records want, and that isn't what they get.

Popular music is recorded on many tape tracks, usually 24 for rock music, and as many as 32 are now commercially available. In recording small groups, a different microphone is put near each instrument, or a signal can come directly from the instrument itself, as from an electric guitar or an electric organ. The voice track may be recorded separately, by a singer who listens to the instrumental recording.

| L | R | L | R | L | R | L | R |

0              Frequency ⟶

■ We can get a pseudostereo signal by feeding alternate frequency bands of a monaural signal to left and right loudspeakers or to left and right headphones. More striking results can be obtained by dividing the monaural source in more complex ways— for example, by using artificial reverberation. Such a system will spread the apparent sound source of an individual instrument out through space.

■ A juke box.

The final two or four tracks are produced by mixing signals derived from the many recorded tracks. Mixing is a complicated process carried out by means of an audio console. Each signal from an individual recorded track is modified in a complex way before the tracks are combined. The modification devices include an attenuator (volume control), a delayer (which can compensate for distance from the microphone), a frequency-response modifier, a timbre modifier (more complex modification of the frequency response), a device to add vibrato and tremolo, a nonlinear distortion producer, a fuzz producer, which introduces high-frequency components of large amplitude and frequency-modulated noise, and a compressor-limiter, which keeps the signal level within a reasonable range. In professional recording, adjustments are changed continually with time, second by second, until a satisfactory overall effect is obtained. Then a final recording is made, with the sequence of time-varying adjustments controlled either by hand or by computer.

What people want changes with time. I can remember when the bass boom of the Majestic radio or juke box was the big thing in music. Something that many people continue to like is a sense of being surrounded by sound. Stereo can do this, and quadraphonic can do it even better.

If we wish, we can be surrounded by the sound coming from a single channel, simply by having different parts of the single-channel sound come from different loudspeakers. Simply having different broad frequency ranges come from different loudspeakers is too crude. Having alternating frequency ranges come from the left and right loudspeakers (or earphones), as shown above, does give a sense of stereo (or pseudostereo) sound.

Better results can be obtained by dividing the sound up in more elaborate ways—for example, artificial reverberation. A reverberant quality can be given to sound by driving a large brass plate with a loudspeaker mechanism, picking up the sound from various points on the plate, and adding this to the original sound. If we drive several loudspeakers with the sounds picked up from several points on the brass plate, we will get a sense of being surrounded by sound, much as in a concert hall, in which the sound reaches our ears reflected from different directions, reflected by different walls.

Reverberation can also be added by placing a loudspeaker and several microphones in a small, hard-walled room. Digital or computer processing of sound is

163

Headphones: surrounded by sound.

a more flexible way to produce artificial reverberation and to divide a sound channel into several parts that differ in complicated ways. Digital reverberation systems are now available commercially.

One can thus in principle get a sense of being surrounded by sound with only one sound channel. What one can't get is an accurate sense of discrete sound sources. An orchestra, a piano, or a singer will seem to surround or envelop us. This is what purists object to.

There *is* one way to "hear it like it is": with two channels and headphones.

With headphones we easily sense the difference between single-channel (monophonic) sound and stereo, but even the stereo sound seems to be inside our head. Ordinarily, we don't externalize such sounds, because they aren't recorded in a way that will allow this. But if the two channels recorded from the ear canals of a dummy head are fed to good headphones through proper filters (to compensate for the fact that the pinna gets into the process twice), the two-channel sound can be heard as external. Tone controls can be used to approximate the proper filters.

Sometimes, dummy-head recordings lead to confusion between sounds in front of us and sounds behind us. Recording from tiny microphones in our own ear canals is better. We hear just what we would in a concert hall, as long as we don't turn our heads. If, in our home, we *really* want to share the same properly-reproduced music among a group of listeners, we must ask them to wear headphones, and all to stare fixedly in one direction. But will *you* hear properly what has been recorded using *my* ears? Alas, we can't try this with stereo as it is now recorded.

High-fidelity enthusiasts often claim that their golden ears diagnose faults that measurements do not disclose. They are much concerned, and rightly, with non-linear distortion, and some, rightly, prefer old vacuum-tube power amplifiers to badly made but elaborate transistor amplifiers. Why? Because the good vacuum-tube amplifiers and the bad transistor amplifiers have different *kinds* of nonlinear distortion.

"Good" and "bad" nonlinear distortion are illustrated at the top of the facing page. In the "good" distortion (part A), the actual curve of output voltage plotted against input voltage departs from a straight line (shown in black)

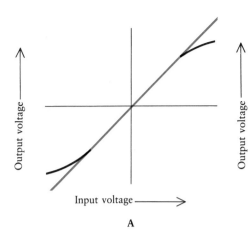

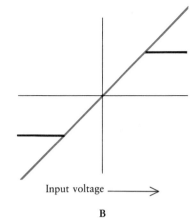

■ "Good" nonlinear distortion (A), contrasted with "bad" nonlinear distortion (B). In each case the curve of output voltage plotted against input voltage departs from the ideal straight line (shown in color). But in the "good" distortion the departure is smooth and gradual; in the "bad" distortion the departure is sudden, which puts a sharp bend in the output waveform, adding high frequencies and giving a distorted sound.

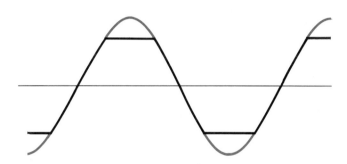

smoothly and gradually. In the "bad" distortion (part B), the curve of output voltage plotted against input voltage bends sharply.

The "good" distortion may actually improve the sound of a solo instrument, making it richer. That is why such nonlinear distortion is sometimes used in processing the sound of a single instrument in recording studios, before the sounds of different instruments are mixed together. Even "good" distortion is bad when the waveform distorted is that of two or more instruments playing different notes, but it may not be very bad.

The "bad" distortion is bad for everything. The figure at the right above shows what it does to a sine wave. It cuts off the smooth peaks and valleys (shown in color) and leaves a squarish wave that sounds harsh and distorted to any ears, golden or not.

There is good reason to prefer vacuum-tube power amplifiers with "good" distortion to transistor amplifiers with "bad" distortion. Some hi-fi enthusiasts go beyond this to resurrect vacuum-tube preamplifiers (which amplify the very low-level signal from a microphone) and use them to replace *good* transistor

■ The effect of "bad" nonlinear distortion on a single sine wave. The bad distortion cuts off the peaks and valleys of the sine curve, giving a squarish wave that sounds harsh and distorted.

■ Harvey Fletcher.

■ Harry Olson at the keyboard of RCA's Electronic Music Synthesizer, 1955.

preamplifiers. I believe they do this because they have been misled by the hi-fi trade, which makes and quotes almost meaningless measurements that disagree with what our ears tell us. But the remedy here should be good measurements, not a refusal to measure at all.

As we noted in the discussion of masking, false signal components that lie close in frequency to strong signal components will be masked, and we won't hear them. However, even a little distortion in a frequency range where there is no signal stands out like a sore thumb. Distortion is commonly measured by using only one or two sinusoidal tones as an input. It would be far more realistic (and difficult) to import a technique for measuring distortion in the broad-band telephone amplifiers used in "carrier" systems to simultaneously amplify many telephone channels. One uses as an input signal a noise signal that contains all the frequencies that the amplifier must amplify *except* in one narrow range of frequencies. With such an input one can measure in the output of the amplifier the distortion power in the range of frequencies that aren't present in the input. This technique has indeed been tried in recent years, but it is not in common use.

In the early days of high-fidelity reproduction, the infant art was in the hands of brilliant and well-equipped scientists and engineers, such as Harvey Fletcher and his colleagues at Bell Laboratories, and Harry F. Olson at RCA, who were recognized leaders in the scientific and engineering community. Somehow, as audio technology has progressed, it has also left the forefront of science and engineering, demoted from the university to the trade school. Perhaps that is proper and inevitable, but it hasn't happened in some European countries as much as it has here. I regret the change, for the good reproduction of sound requires deep insight as well as good technology.

Today the hi-fi community faces a fresh problem for which it is ill-prepared: computer-produced sounds. Here there is no original sound to reproduce. A sequence of computer-generated numbers is converted into an electric signal by a digital-to-analog converter. The electric signal from the digital-to-analog converter is amplified and drives loudspeakers. The very first time there *is* a sound is when a sound wave comes from the loudspeakers. That sound may be tailored to sound best in an auditorium, large or small, or in your very own living room. Distortion in amplifiers and speakers would have to be defined in terms of the composer's intentions, since whatever the speakers produce is the "original" sound.

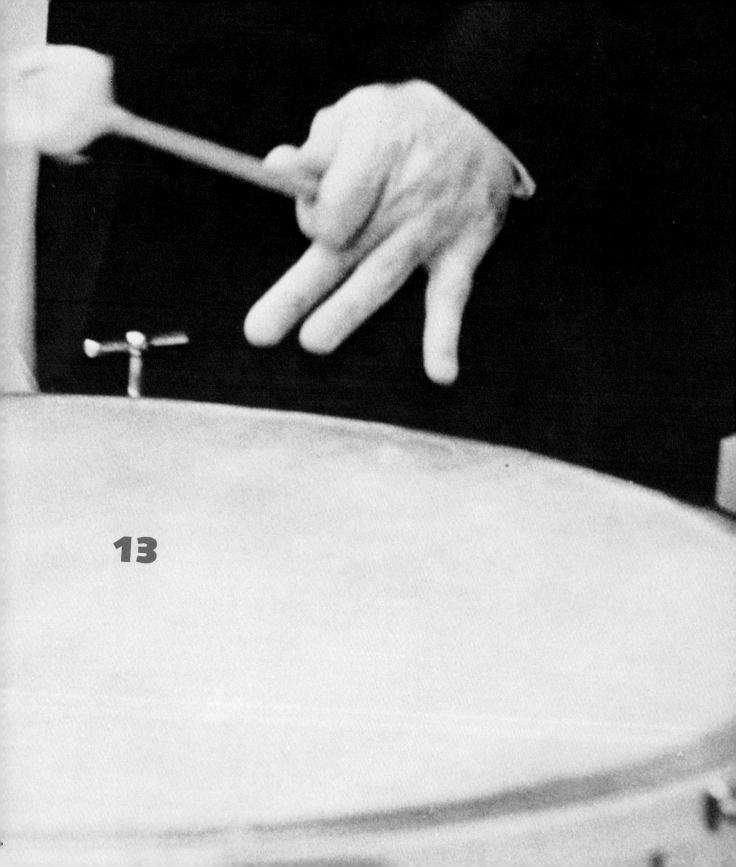

13

# Analysis, Synthesis, and Timbre

**13**

Pitch and loudness are straightforward compared with the puzzle of timbre. For periodic nonsinusoidal musical sounds, pitch is tied inextricably to periodicity; we respond to 440 periods a second with the sensation of A above middle C. Loudness is tied to intensity, though we have learned that the loudness produced by the sinusoidal components of a sound depends on their frequencies and on whether they lie within a critical bandwidth.

Timbre is much more complicated. Surely, it must depend in some way on the spectrum of the sound; yet musicians can recognize different instruments played over a pocket-sized transistor radio that seriously distorts the spectra of sounds, cutting off many low frequencies entirely. A saxophone still remains a saxophone when heard over such a radio, an oboe remains an oboe, a bassoon a bassoon, a violin a violin, and all are distinguishable from a French horn or the human voice.

The puzzle is great, for musicians can recognize musical instruments played over a wide range of pitch. Perhaps they learn to associate with one instrument the qualities of the sounds in different pitch ranges. But even to the casual listener the different pitches of the same instrument have a good deal in common. And what about loudness? A *forte* sound is different from a *pianissimo* sound in more than intensity, just as a shout differs from a quiet voice. Merely turning the volume control up or down won't change one sound into the other.

Our understanding of timbre is still incomplete, but much has been learned in recent years, both from the analysis of sound waves—that is, by examining their waveforms and spectra—and from the synthesis of sound waves—that is, by producing (usually by means of a computer) combinations of sine waves of prescribed spectra and time variation. Analysis tells us what the waveform and spectrum of a musical sound are. It shows complicated features that may or may not be important in our perception of the sound. It is only by synthesizing a sound, incorporating some features revealed by analysis, simplifying or disregarding others, that we can find what features of waveform and spectrum are important in producing what effects.

A recognition of the importance of both analysis and synthesis in the study of musical sounds goes back to the work of Helmholtz. His analytical tools were crude by our standards. He could watch the path traced out by a bright speck on a violin string or a tuning fork. He could seek out partials by listening through a

number of glass resonators. He could listen for partials carefully, focusing his attention by means of a tone from a tuning fork or a musical instrument. And he could calculate the relative strengths of the partials that are produced when a stretched string is plucked or struck.

Helmholtz's means for synthesizing sounds were even cruder. He could listen to the sounds of several tuning forks vibrating simultaneously. He could listen to sounds produced by sirens with several different rings of holes. That was about all.

Only those who have looked through Helmholtz's *Sensation of Tone* can appreciate the complexity and variety of his work. Two simple outcomes have been all but unquestioned before our own day of easy and exact synthesis of sounds by means of computers. The first was a casual belief that musical timbre depends only on the relative amplitudes of the partials present in a sound. The second was that the relative phases of those partials are unimportant to the ear. We will consider the matter of phase first.

Plomp and others, by very careful listening under laboratory conditions, have shown that one can detect changes in the relative phases of a tone made up of a first partial and a second (octave) partial. This means that one can hear a beat between pure tones that are an octave apart in frequency. Such beats are heard as changes in sound quality rather than in sound intensity, and can be heard between pure tones that are close to a consonant interval in frequency, such as a fifth or a fourth. One must listen very carefully to sense such changes in sound quality with the relative phases of pure tones. The effect is not musically important.

**A**

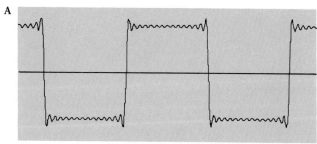

**B**

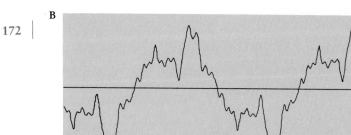

The waveform depends on the relative phases of the partials as well as on their relative amplitudes. Part **A** shows an approximation to a square wave made up of 12 partials with the "correct" phases. Part **B** shows a wave made up of partials with the same amplitudes but with randomized phases. The waveforms are very different. The pitches remain the same. The sound of the wave in part **B** is perhaps a little "mushier" than that of the square wave (in **A**).

Suppose that we make a tone up of many (say, 10 to 20) harmonic partials, so chosen that together they approximate a square wave. If the phases of the partials are "correct" the waveform will indeed be a square wave, as shown at the top of page 43, where we first encountered the representation of a square waveform by sinusoidal components. Such a square wave sounds harsh and raspy.

Suppose that we randomize the phases of the same partials that, with correct phases, give a square waveform. Part A of the above figure shows the square wave made up of 12 partials, and part B shows a wave made up of partials with the same amplitudes but with random phases. This waveform certainly *looks* different. Does it *sound* different? It isn't easy to tell. Perhaps it sounds a little mushy, and less raspy than a square wave. The pitch is unaltered.

Such phase effects are easier to hear with headphones than in a reverberant room. Phase has some importance for musical sound, but not as much as some purists insist. However, phase does show up strikingly in the sound of the piano.

We noted in Chapter 3 that the partials of the piano strings are not quite harmonic. The stiffness and tension of the strings tend to keep them straight. The effect of the stiffness is greatest for high partials, for which the vibrating string has many short bends. The extra force added by the stiffness of the string makes such partial frequencies appreciably higher than they would be for a stretched string without stiffness. The degree of this inharmonicity is such that the fifteenth partial can have 16 times the frequency of the fundamental.

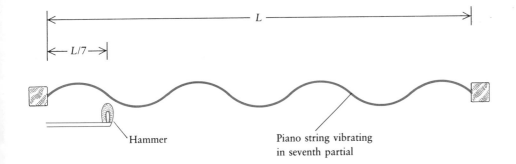

$L$

$L/7$

Hammer

Piano string vibrating
in seventh partial

■ In some pianos, the hammers are placed about a seventh of the string's length from the end of the string, so that the hammer cannot excite a vibration of the string at the frequency of the seventh harmonic. Plucking the string a seventh of the way along its length could not excite the seventh harmonic, because in this harmonic mode the string does not move at a point a seventh of the way along its length. The action of a piano hammer is more complex than plucking, because of the size of the hammer and the manner and duration of its contact with the string; so it may partially excite the seventh harmonic even if it strikes a seventh of the way along the string. The hammer would be very effective in exciting the seventh harmonic if it struck a fourteenth of the distance along the string, where the motion of the seventh harmonic is greatest. The intensities of other harmonics also depend on where the hammer strikes the string.

In 1962 Harvey Fletcher and his collaborators studied the sound of the piano by means of analysis and synthesis. They found that the warmth of the piano tone depends entirely on the *inharmonicity* of the partials. How did they discover this? By synthesizing sounds that were close approximations to piano sounds in the rise and decay of all partials, but for which the partials were exactly harmonic. Such sounds were bland and uninteresting. They had none of the wavering quality that makes a piano tone warm. The wavering quality is caused by the continual shifting of phases, or beating, of the higher partials, from the continual change of waveform that results because the partials are not exactly harmonic.

The quality of the piano tone must in some degree depend on what partials are actually excited when the hammer strikes the string. It is a commonplace of piano lore that the hammers are placed to strike at a point one-seventh of the way along the string (see the above drawing), so that the hammer cannot set the string vibrating in the pattern of the seventh partial, shown in the drawing as a displacement of the string. Traditional harmony (and the diatonic scale) are based on relations among the first six harmonic partials. The seventh is a noteworthy intruder, as are many higher partials, whatever good color they may lend the tone.

The hammer position affects what partials other than the seventh will be excited, and in what relative intensities. The nature of the hammer is important, too. A hard hammer excites more higher partials and gives a bright tone. Softening the hammer by pricking the felt repeatedly with a pin weakens the higher partials and gives a mellower tone.

The sound of a harpsichord is less soft than the sound of a piano. In the harpsichord the string is plucked by a small sharp quill. This sets up many high-frequency partials that give the harpsichord its jangly sound.

The piano, the harpsichord, the guitar, and certain other string instruments have a characteristic plucked or struck sound. This can be verified by computer synthesis of sounds. From the earliest days of computer sound synthesis, it was observed that any abrupt rise coupled with any gradual decay gave such a sound, whatever the shape of the rise or decay, and regardless of the waveform. Some of the differences between timbres must be associated with the attack and

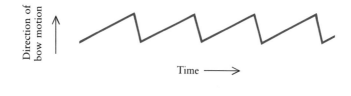

Direction of bow motion →

Time ⟶

■ The motion of a bowed violin string has a sort of sawtooth pattern of displacement with time. The bow drags the string along until a reflected wave from the *nut* (the small ridge at the peg-end of the violin over which the strings pass) causes the string to slip past the bow. The bow catches the string again and pulls it along, as indicated in this figure.

174

■ Max Matthews's electric violin.

decay of the sounds, that is, with how their overall intensities (or the intensities of their individual partials) rise and fall with time.

Nonetheless, relative strengths of the partials do affect the quality of a sound. They help us distinguish the sound of the harpsichord from the sound of the piano. The somewhat harsh quality of the jangly harpsichord is characteristic, as we have noted earlier, of sounds for which many partials lie within one critical bandwidth.

This is the explanation of the unpleasantly buzzy sound of electronically generated square waves and sawtooth waves.

As observed by Helmholtz and confirmed by Mathews and others, the motion of the violin string while it is bowed follows a sawtooth pattern, as shown in the figure above. The bow drags the string along for a short distance, then the string slips back, only to be picked up and again carried along with the motion of the bow. If this were the waveform of the sound of the violin, that sound would be harsh indeed. However the soundboard of the violin has many resonances, which reinforce some partials and suppress others. The figure on the facing page shows a plot against frequency of the effectiveness of the violin body in transforming a partial of the string's vibration at the bridge into a partial of the sound wave that the violin produces. In the sound of the violin, a few partials whose frequencies lie close to the resonances of the violin are much more intense than those nearby. Jean-Claude Risset has noted that this has an interesting and important connection with vibrato. Even this small variation of frequency causes amplitude modulation of the harmonics of the tone, because the vibrato shifts the frequencies of the harmonics toward or away from the resonant peaks shown in the figure on the facing page. This effect is very apparent in some recordings of the waveforms of violin tones that Mark Dolson made recently at the California Institute of Technology. The waveform changed markedly during the vibrato, showing that the relative amplitudes of different harmonics do indeed change as the fundamental frequency of vibration of the string shifts up and down.

Max Mathews has made an electronic violin in which the sawtooth motion of the string is converted into an electric signal by means of a magnetic pickup. By passing this signal through an electrical network having from 17 to 37 resonances that matched those of a good violin, he obtained a very good string

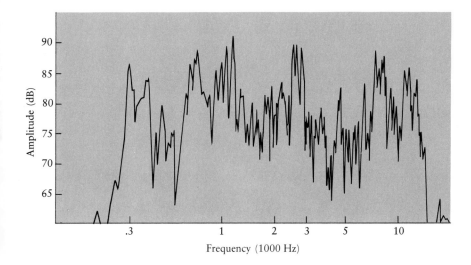

■ A sawtooth wave, such as that in the figure on the facing page, has a very harsh sound. What we hear is not the motion of the string, but that motion translated into a sound wave in the air through the soundboard (and other structures) of the violin. The soundboard has many resonances. The curve in the figure above shows how effectively a partial in the motion of the string is transferred to a partial in the sound wave radiated by the violin. Among the higher-frequency partials, some will lie near a resonant frequency of the violin and produce strong sounds; others will lie in the valleys of the curve and will produce little sound. This is very important to the sound quality of a violin. Much of the difference between good and bad violins depends on the location of the peaks and valleys of such a curve.

quality. (He later obtained a good string quality without such a filter. Perhaps the somewhat unsophisticated loudspeakers that he uses have jagged resonances like those of the wooden structure of an actual violin.)

In order to obtain a good sound in playing double stops, Mathews found that he had to use a separate amplifier and loudspeaker for each string. When a single amplifier and loudspeaker were used for all four strings, small nonlinearities caused the signals from the strings sounded simultaneously to produce unpleasant combination tones whose intensities, though small, were great enough to be annoying.

Furthermore, by using a low-pass filter whose bandwidth increased with intensity, Mathews was able to make his electronic violin produce a convincing brass sound. This accorded with Risset's observation that in trumpet tones the higher partials appear later than the lower partials. By using a band-pass filter whose center frequency rose with amplitude, Mathews made the electronic violin produce a voicelike "wah" sound. More recently, he has used a filter that adds to the sound of a violin the "singer's formant," with a frequency of about 3,500 Hz, that was discussed in Chapter 9 in connection with masking. This gives the electronic violin a pleasing sort of super-resonant sound. Alas, a string section using such electronic violins would mask a singer's voice!

In the violin the frequencies that are emphasized are determined by the resonances of the soundboard, not by the frequency of the note played. If a resonance at 2,200 Hz is excited by the tenth harmonic of a 220-Hz note, it will also be excited by the fifth harmonic of a 440-Hz note. In other instruments, tone color depends on the relative intensities of particular harmonics. For example closed organ pipes have a "hollow" sound because only odd harmonics are present; open pipes have a fuller sound because they include both even and odd harmonics.

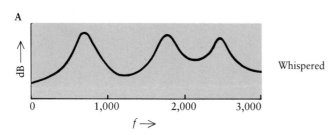

A

Whispered

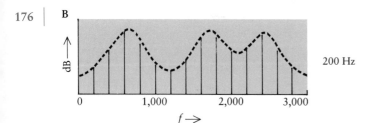

B

200 Hz

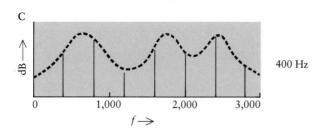

C

400 Hz

■ Vowel sounds are character-ized by resonances of the vocal tracts, or *formants*, which make the speech sound more intense in narrow frequency ranges. The curves shown in this figure illus-trate the effect of these reso-nances for the vowel *a* as in *had*. Curve **A** illustrates a whispered *had*. This is a noiselike sound, and a whole range of frequencies are present. *Had* spoken with a pitch frequency of 200 Hz is il-lustrated in part **B**. Only partials of frequencies 200 Hz and its harmonics are present. The dashed *envelope* curve indicates the effects of the resonances of the vocal tract on the intensities of these partials. In part **C**, the pitch frequency is 400 Hz. There are only half as many partials as in part **B**. One might guess that it is harder to find the formants from a recording of a female voice than from a recording of a male voice, and this is true. The widely spaced partials of the fe-male voice do not indicate the formant frequencies very clearly.

The vowel sounds of the human voice are distinguished and perceived, regard-less of pitch, because of the three chief resonances or *formants* of the vocal tract. Near the formant frequencies the intensities of the harmonics of the sound pro-duced by the vocal cords are strong; harmonics far removed from the formant frequencies are weak. Indeed, we can distinguish the vowels in whispered speech, in which all frequencies are present rather than a sequence of harmon-ics. In part A of the above figure, we see the spectrum of a whispered *a* as in *had*. In part B, we see the same vowel with a pitch of 200 Hz. In part C, we see the same vowel with a pitch of 400 Hz. In parts B and C the envelope, or outline, of the peaks of the spectrum, which represents the resonances of the vocal tract, is roughly the same as in the spectrum of the whispered vowel.

I have said that the frequencies of the formants characterize a vowel sound, regardless of pitch. This is not quite true for the singing voice, as Johann Sundberg has shown. For one thing, the singer manages to produce a high *sing-*

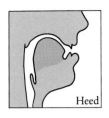
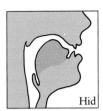
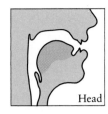
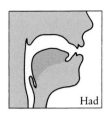

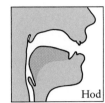
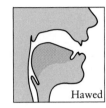
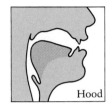
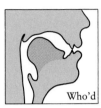

■ Vocal tract positions for some English vowels. The vowels in *heed*, *hid*, *head*, and *had* are called *front vowels*, because the highest point of the tongue is in the front of the mouth. The vowels in *hod*, *hawed*, *hood* and *who'd* are called *back vowels*, because the position of the tongue is in the back of the mouth. The tongue is highest in the vowels of *heed* and *who'd*, which are called *high* or *close* vowels, and lowest in the vowels of *had* and *hod*, which are called *low* or *open* vowels. As to timbre, the vowel of *heed* seems shrill, and the timbre of the vowel in *who'd* seems low or dull.

*er's formant* that makes the singing voice intense in a frequency range in which common orchestral sounds have little power. Furthermore, sopranos shift their formants when they sing very high notes. This both makes the sound louder and alters its quality. Computer synthesis of the singing voice shows that, if this is not done, the voice sounds like that of a child, not a woman. Finally, one can control the formants or resonant frequencies of the vocal tract so that they coincide with particular harmonics. One can seem to sing several notes at once, though all are, of course, harmonics of the frequency of vibration of the vocal cords. The accurate control of formant frequencies gives a wonderful quality, not found in ordinary speech or singing, to some Buddhist chanting.

The formant frequencies are clearly an extremely important aspect of the human voice, and are worthy of our attention. The figure above shows vocal-tract configurations, and the table below lists the corresponding formant frequencies, for common English vowels.

We might be led to conclude that frequency spectrum is all there is to vowel sounds. Try saying *ah* as a very prolonged monotone, without any vibrato, without any change in pitch or intensity. I think you will find that the sound gradually loses its *ah* character. It becomes a rather buzzy, not too pleasant tone, without much character. What happened to the *ah*ness?

Formant frequencies of common vowels.

|        | Heed  | Hid   | Head  | Had   | Hod   | Hawed | Hood  | Who'd | Hud   | Heard |
|--------|-------|-------|-------|-------|-------|-------|-------|-------|-------|-------|
| $f_1$  | 270   | 390   | 530   | 660   | 730   | 570   | 440   | 300   | 640   | 490   |
| $f_2$  | 2,290 | 1,990 | 1,840 | 1,720 | 1,090 | 840   | 1,020 | 870   | 1,190 | 1,350 |
| $f_3$  | 3,010 | 2,550 | 2,480 | 2,410 | 2,440 | 2,410 | 2,240 | 2,240 | 2,390 | 1,690 |

Psychologists have a label for this: *semantic satiation*. If we utter the same word over and over again, we come to hear it simply as a sound. The meaning has been conveyed at the first hearing, and it is not indefinitely reinforced by repetition. We get no more meaning; we become satiated.

When we hear a vowel sound, our first impression is to recognize what vowel we hear. As the vowel is prolonged, we come to hear it simply as a sound, perhaps because much of our nervous system responds to changes rather than to continued stimuli. We sense the onset and the first parts of sounds differently from their later parts. And, throughout the sound, changes are welcome. The rise and fall of sound, vibrato, its onset and diminution—all are important to the ear. Indeed, in a passage of music it is important that successive notes don't sound *exactly* the same.

Perhaps the first thorough experiments that demonstrated unequivocally the importance of change and variety in sound were those carried out by Jean-Claude Risset at Bell Laboratories and published in 1966 and 1969. Risset used a computer to analyze in great detail the rise, fall, and variation with time of the various partials of recordings of short trumpet tones played by a professional trumpet player. He found the sounds to be exceedingly complex. By using a computer to synthesize sounds that matched selected fine details of the analyzed sounds, he found that some complex features of the real trumpet sounds were important to the ear, and some weren't. For example, short-term fluctuations of the amplitudes of various partials turned out not to be important to the ear. Neither was a short burst of noise found at the beginning of real trumpet tones.

What was important? It proved important that the higher partials start later and fall earlier than the lower partials. Although random variations in amplitude of the partials proved unimportant to the ear, random variations of the frequencies of the partials were important in giving synthesized sounds the brassy sound of a real trumpet. Omission of all frequencies above 4,000 Hz, from either real or synthesized sounds, did not make them less recognizable as being trumpet sounds, but did make them seem less brilliant. When trumpet sounds were synthesized with reasonable attention to the rise and fall of the intensities of the various partials, and with appropriate random frequency changes or vibratos of various partials, trumpet players could not tell the synthesized short trumpet sounds from real short trumpet sounds.

■ Jean-Claude Risset.

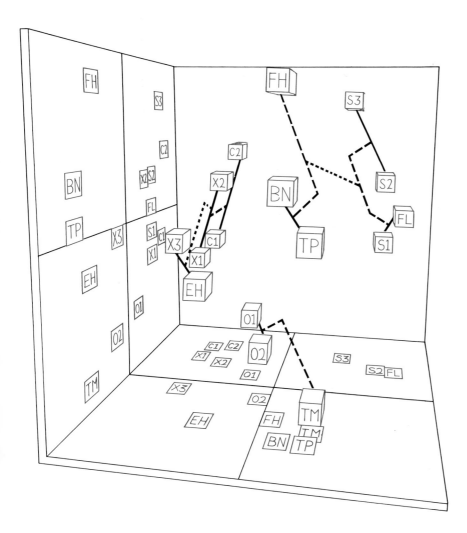

■ Three-dimensional display of differences and similarities between instrumental sounds based on numerical ratings of similarity or dissimilarity for various pairs of sounds. The display was obtained by a computer technique called *multidimensional scaling.* In it, instrumental sounds that were judged to be similar are close together, those judged dissimilar are far apart. The dotted and dashed lines that connect the members of several groups of instruments were obtained by a different technique, called *clustering.* The squares on the walls are two-dimentional projections of the cubes, showing their positions up-and-down and forward-and-back (left wall), and left-and-right and forward-and-back (bottom wall). The abbreviations are: O1, O2, oboes; C1, C2, clarinets; X1, X2, X3, saxophones; EH, English horn; FH, French horn; S1, S2, S3, strings: TP, trumpet; TM, trombone; FL, flute; BN, bassoon.

We noted earlier that turning up the volume control of an amplifier does not transform a soft utterance into a shout. Similarly, the louder a trumpet tone is, the greater the fraction of the energy that is in the higher partials.

In his early work, Risset synthesized only individual, short trumpet tones. Since then, Dexter Morrill of Colgate University has synthesized very convincing trumpet passages. Furthermore, the sounds of many musical instruments have been studied in detail, though usually for tones of one pitch only. At Stanford, John M. Grey used a computer to analyze and synthesize various instrumental sounds, in order to identify and eliminate features of real musical sounds that make little or no difference to the trained ear. He then synthesized instrumental sounds, equal in duration, loudness, and pitch, which proved to be difficult or impossible to distinguish from sounds produced by the instruments they imitated.

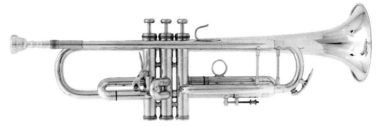

■ A B-flat trumpet.

■ An E-flat clarinet.

180 Grey also asked trained musicians to rate the similarity of pairs of musical sounds. The rating scale was *very dissimilar*, 1 to 10; *dissimilar*, 11 to 20; *very similar*, 21 to 30. He sought to get a simple representation of the similarities of these sounds by means of a computer technique called *multidimensional scaling*. This gave him the three-dimensional representation in the diagram on the preceding page. In this representation, each cube stands for a particular instrument, and distance between cubes corresponds to their similarity.

On this basis of similarity, the instruments fall into three families, each with several subfamilies. These are as follows.

**1.** E♭ clarinet (C1); soprano saxophone, *mf* (X1); soprano saxophone, *f* (X3); bass clarinet (C2); soprano saxophone, *p* (X2); English horn (EH).

**2.** oboe (O1); muted trombone (TM).

**3.** bassoon (BN); French horn (FH); cello, *sul ponticello* (S1); cello, normal (S2); trumpet (TP); flute (FL); cello, *sul tasto* (S3).

Grey next looked for the physical characteristics of the sounds responsible for these similarities. He found that up-and-down (I) can be interpreted as *spectral energy distribution*. Near the top (FH and S3), the spectrum is narrow and has its peak at a comparatively low frequency. Near the bottom (TM), the spectrum is wide and peaks at a higher frequency. TP and X3 lie in the midrange.

Left-and-right (II) seems to depend on whether the partials rise and fall at the same time. In the sounds of the woodwinds (left), the various partials do rise and fall at the same time. This is not so for the strings, flute, or brass, nor for the bassoon, which is a strange woodwind.

Forward-and-back (III) seems related to the initiation of the sound. In the sound of the flute (FL) and strings (S1, S2, S3), a short burst of noise precedes the tone and is important for its quality. This is not so important for the trumpet (TP), the trombone (TM), or the bassoon (BN). However, too large a burst of noise at the beginning of a synthesized violin sound gives the effect of an inexpert player.

It is not entirely clear precisely what is represented by the up-and-down spectral energy distribution dimension, I. Perhaps this is a sort of average of the frequencies of the partials, weighted by their computed loudnesses.

It is clear that many physical aspects of musical sounds, including spectral infor-

■ A B-flat soprano saxophone.

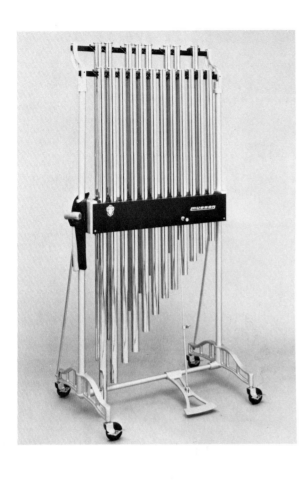

Partials of a chime.

| 1st | 2nd | 3rd | 4th | 5th | 6th | 7th |
|-----|-----|-----|-----|-----|-----|-----|
| $f_0$ | $2.76\,f_0$ | $5.40\,f_0$ | $8.93\,f_0$ | $13.34\,f_0$ | $18.64\,f_0$ | $31.87\,f_0$ |
| — | — | $(4.5\,f_0)$ | $2 \times 4.47\,f_0$ | $3 \times 4.45\,f_0$ | $4 \times 4.66\,f_0$ | $7 \times 4.55\,f_0$ |

■ Orchestral chimes.

mation, contribute to their timbres. Spectral information enables us to distinguish the various vowel sounds. The relative times and rates at which various partials rise is important in brass sounds. An initial high-frequency burst of noise is essential to the timbre of some instruments, such as flutes and strings. And we should remember that a fast attack and a gradual decay gives the effect of a plucked or struck string.

What about bells and gongs, which are also struck? Bells and gongs differ from strings in that, as we have seen, their partials are not harmonically related. Their tones do not give conventional effects of consonance and dissonance when we use them to play conventional harmony; yet we can play recognizable melodies on bells and gongs, which do give a sense of pitch.

A great deal of study has been devoted to bells, gongs, and related instruments. Among these are orchestral chimes, which are long, uniform metal tubes hanging loosely from one end so that they can flex and vibrate freely. The frequencies of the first few partials of a typical chime are shown in the table above. Here the fourth, fifth, sixth, and seventh partials are approximate harmonics of a frequency equal to $4.5f_0$, which is therefore heard as the perceived pitch of the chime. This is a manifestation of Schouten's residue pitch (see Chapter 6).

182

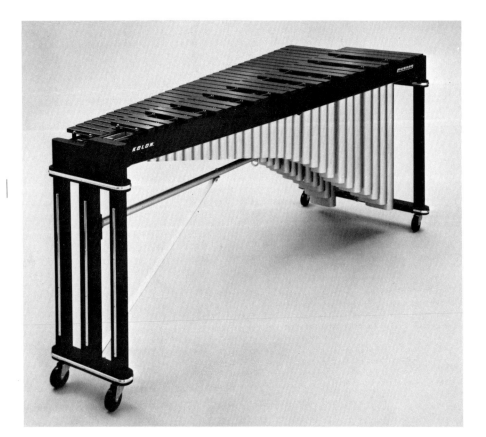

■ A marimba.

■ A xylophone.

A thin board bends more easily than a thick board. Imagine that the tube of a chime is not exactly round, but is flattened a little. It will bend more readily in the flattened direction than in the thickened direction. A chime that is very slightly out of round will have two sets of partials, one with slightly higher frequencies than the other. These partials will beat, giving a slightly wavering sound. This beating or wavering is disliked in chimes, but it is an attractive feature of the sound of gongs. Such wavering has been used to good effect in computer-synthesized bell and gong tones.

Orchestral bells, or the glockenspiel, give high-pitched sounds that are scored (in musical notation) two octaves lower than they sound. Orchestral bells consist of loosely mounted metal bars. These have many high partials, some corresponding to bending and some to twisting motions. The high partials die away rapidly after the bar is struck, and the perceived pitch is the frequency of the first, bending partial.

Marimbas and xylophones use wooden or synthetic bars, thinner in the middle than at the ends. This thinning of the middle of the bar causes the second partial of the marimba bar to lie two octaves above the first partial. The xylophone bar is thinned less, and the second partial has a frequency three times that of the first partial. Closed tubular resonators under the bars of marimbas and xylophones increase the sound intensity of the first partial and make it die out more quickly. In the xylophone, the closed tube under a bar has another resonance, at three times the frequency of the fundamental—that is, at the frequency of the second partial of the xylophone bar—and so strengthens its intensity also. This is one reason why the xylophone has a brighter quality than the marimba. However, the xylophone sounds brighter because it is commonly played with hard mallets, whereas the marimba is more often played with soft mallets.

The vibraphone, vibraharp, or vibes has aluminum bars shaped much like those of the marimba. The vibrations of these bars decay more slowly than those of the marimba. Motor-driven disks between the tops of the resonators and the bars above them alternately open and close a passage between bars and resonators, and give the vibraphone its wavering quality.

The phenomenon of residue pitch also explains the perceived pitch of bells, whose modes of oscillation are very complex. By careful casting and shaping, Dutch bell founders of the seventeenth century, especially the Hemony brothers,

A Hemony bell.

managed to produce bells whose lower partials had the orderly relation shown in the table on the facing page. The sound of such a bell is more like a complex chord than a usual musical tone. In ordinary bells, the frequencies of the partials depart from these orderly values and many other partials are present as well; yet we hear the bell as having a characteristic, unified sound and a definite pitch. Bells of different sizes always differ in their perceived pitch, however complex their sounds may be. By striking gongs or bells of different sizes we can play a tune, but we can't make conventional harmony.

Kettledrums, or tympani, have many partials, but the pitch is that of a mode of vibration called the *principal tone*. There are several partials present whose frequencies are about 2, 3, 4, and 5 times *half* the frequency of the principal tone, but for some unexplained reason the pitch of the kettledrum corresponds to the frequency of the principal tone, instead of being an octave lower. The Indians have ingeniously made nonuniform drumheads that vibrate with almost harmonic frequencies and give a clear sense of pitch. But most drums, such as bass drums, snare drums, tomtoms, conga, or bongos, have no clear pitch, and are used solely for their percussive effect.

In Chapter 6, I described computer-generated sounds with nonharmonic partials that didn't "hang together." The upper partials didn't fuse with the lower partials to give a sense of a single musical sound. Nevertheless we hear bells, gongs, drums, and even wood knocking on wood, as in castanets, as single, distinguishable sounds, even though their partials are nonharmonic. In part, a sharp attack followed by a slow decay helps to make sounds hang together. Elizabeth Cohen found this to be true for stretched tones, unless they were too strongly stretched. Perhaps we identify some complex natural sounds because we have heard them many times, and have learned to recognize and name the physical source. Certainly, by careful listening we can hear various frequency components of sounds such as those of chimes, bells, and gongs; yet this does not keep us from identifying the whole sound as one sound, characteristic of that particular instrument. It may be that we have not yet identified some subtle characteristic by which we are able to recognize certain kinds of sound.

Partials of tuned bells.

| Partial | Relation to $f_p$, the perceived pitch |
|---------|----------------------------------------|
| Hum tone | $0.5\, f_p$ (an octave down) |
| Prime | $f_p$ |
| Third | $1.2\, f_p$ (a minor third up) |
| Fifth | $1.5\, f_p$ (a fifth up) |
| Octave | $2\, f_p$ (an octave up) |
| Upper third | $2.5\, f_p$ (an octave plus a major third up) |
| Upper fifth | $3\, f_p$ (an octave plus a fifth up) |

■ An Indian kettledrum

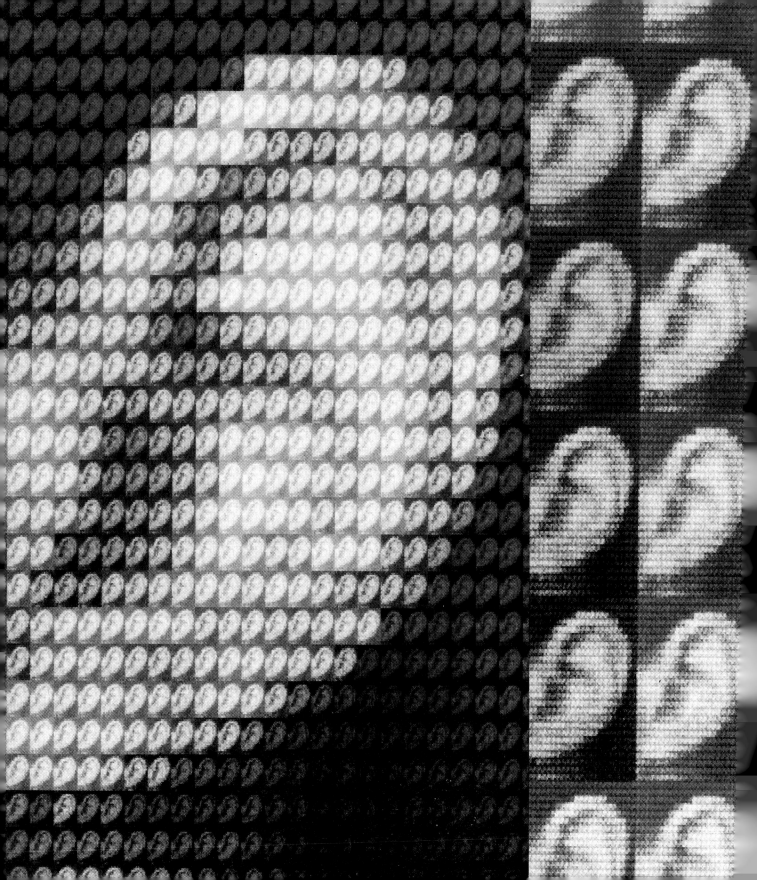

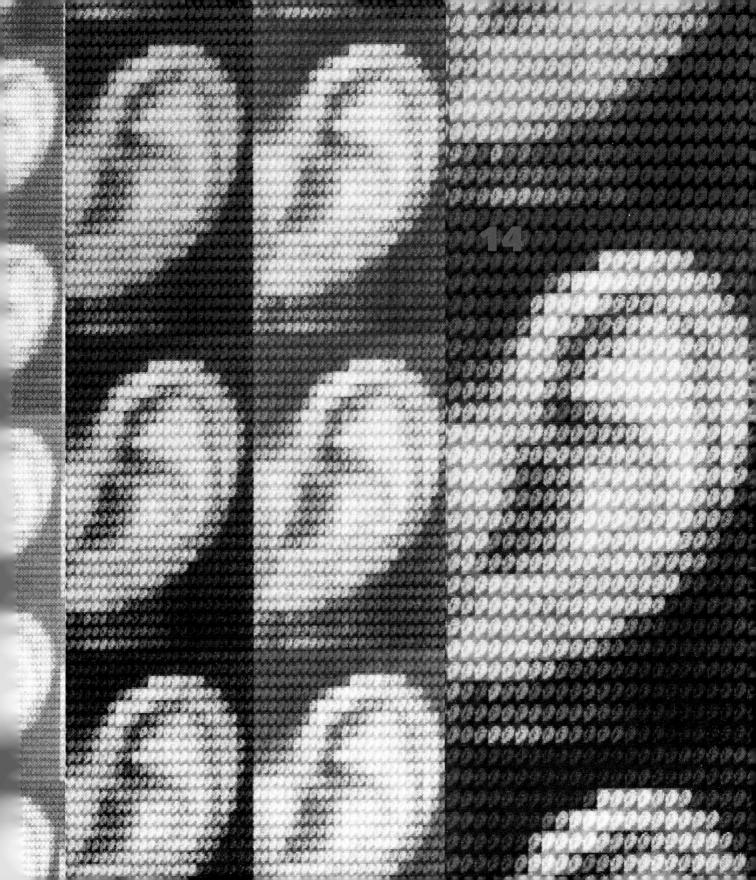

# Perception, Illusion, and Effect

**14**

The human organism, ill-knowing and ill-understood, confronts itself across a perplexing territory of which there is only a rudimentary map. A listener is endowed with, and bound by, abilities of perception, keen to detect some subtleties, deaf to others, and open to deceit and illusion. The traditional composer has a canny knowledge of the sounds that players can evoke from instruments and of how these sounds can be made to contrast or blend. A composer of computer sounds knows something of how and why the paper cone of a loudspeaker can ring like a bell, blare like a trumpet, speak with a human voice, or produce sounds and illusions never before heard by human beings.

Our constitution, our mind, and our senses are central to the overall process of the production and perception of music. There is much that we do not know. Indeed, recent discoveries make us appreciate the depth of our ignorance.

Medical literature gives curious accounts of the *amusias*, defects of musical ability associated with disease or injury of the brain. These include loss of ability to sing (without words), loss of ability to write musical notation, loss of ability to recognize familiar melodies, and loss of ability to read musical notation. There have been studies of the impairment of more detailed skills, such as the correct perception of temporal order, of simultaneity, of duration, and of rhythm. Some of the deficiencies are associated with impaired function of the left or "dominant" hemisphere of the brain, where, in right-handed people, speech and the understanding of language reside. But musical deficiencies and speech disorders or *aphasias* are not inevitably linked.

Although right-handed people who have suffered extensive damage to the left hemisphere of the brain may be able to speak and understand only with great disability, or not at all, work by Roger Sperry and his followers on split-brain patients, in whom the two hemispheres have been surgically isolated, has broadened and refined our understanding of these and other abilities. The right hemisphere can exhibit rudimentary language capabilities. It is better than the left at solving difficult geometric puzzles. The relative predominance of various abilities in the two hemispheres is clearly a fact, but not one yet completely understood.

What about music? There has been a good deal of controversy. It has been known for many years that injuries to the left hemisphere *can* result in loss of ability to sing or whistle a tune; yet in 1966 a patient whose entire left hemisphere had been removed, and who had lost the ability to speak, was found able to sing familiar songs with few articulatory errors.

Such puzzling phenomena probably indicate that a complex combination of abilities is needed for the performance and perception of music. Recently, it has become possible to locate the site of mental activities by injecting into a subject's arteries a radioactive substance that is absorbed by the part of the brain engaged in a task. Several subjects were asked to report whether two groups of tones were different or similar. In some subjects, who tried to remember the succession of tones as a heard melody, the right hemisphere was more active during the task. In other subjects, who mentally plotted the tones on a musical staff, the left hemisphere was more active. Here was one simple musical task. Different people approached it differently, with different parts of the brain.

What is the significance of this for musical sound? We don't really know, except that the process of musical perception can be very complex and can differ from person to person.

Musical illusions are interesting in themselves and can shed light on our complex powers of perception. In the October 1975 issue of *Scientific American,* Diana Deutsch published an interesting article on musical illusions. One of these illusions is represented below. As part A shows, a sequence of paired tones was presented to the right and left ears: first high to the right, low to the left; then

Stimulus to right ear

Stimulus to left ear

Heard by right ear

Heard by left ear

■  A musical illusion described by Diana Deutsch. The stimuli presented to the left and right ears through headphones are shown in part **A**, and what the subject heard is shown in part **B**. Although both high notes and low notes were presented to each ear, the right ear heard only high notes, the left ear only low notes. (Some people do not experience this illusion.)

low to the right, high to the left; and so on. As part B shows, a right-handed listener commonly heard a high tone in the right ear, followed by a low tone in the left ear, then a high tone in the right ear, and so on. Not only did the right ear disregard the low tones, and the left ear the high tones, but the left ear "heard" a low tone that was present only to the right ear. Deutsch explained this by saying that only one ear (in this case, the right) perceived pitch and that localization was perceived by a separate mechanism which homed in on the higher pitch. Another illusion described by Deutsch, one that may seem congenial to musicians because, in a sense, it brings order out of seeming chaos, is shown in the figure below.

Stimulus to right ear

Stimulus to left ear

Heard by right ear

Heard by left ear

■ Another musical illusion described by Diana Deutsch. The tones presented to the left and right ears, in part **A**, make up an ascending scale and a descending scale. What was heard is shown in part **B**. The right ear heard the higher tones as running part way down the scale and then up again. The left ear heard the lower tones as running part way up the scale and then down again.

Whatever the explanation of such phenomena of perception, they should warn the computer musician to beware. However, such binaural presentations of sounds are unusual in music. Other peculiar effects occur more commonly. These can be studied most easily by means of computer-generated sounds, whose qualities, including timbre, can be specified and varied with great accuracy.

The figure on the facing page illustrates an experiment described in 1978 by David L. Wessel. Staff A shows the pitches of successive notes. The timbres of the notes marked + and × can be different.

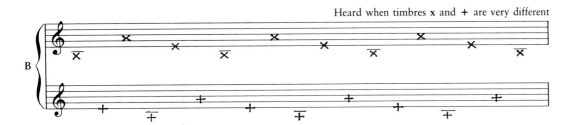

When all notes were the same, or only slightly different, in timbre, the succession of notes was heard as written in part A; that is, as a repeated pattern of three ascending notes. However, as the difference in timbre between + and × increased, the notes were heard as two separate voices (as shown in part B), each voice a repeated pattern of three descending tones. Extensive studies of such *streaming* effects by Albert S. Bregman, Wessel, and others show that notes can be bound together in a stream by differences in spectrum, but not by differences in attack.

This powerful effect of timbre that Wessel found led him to suggest that timbre could be used as a musical "dimension," like pitch or loudness. His work, and earlier work of Grey, make it possible to map at least some timbres in an orderly way, as in the three-dimensional timbre space of the diagram on page 179. Can we make music out of an organized progression through such a space? Timbral transitions have been used effectively in Chowning's *Turenas* and *Phōnē*, and in several compositions by Jean-Claude Risset, of which examples are given on the cassette recording that can be ordered from the publisher of this book.

The trick shown in the figure above, producing more than one voice by playing successive notes with different timbres, is not uncommon in traditional music. It is used to particularly good effect in Bach's works for the unaccompanied violin (for example, see the score on the next page). Separation in pitch helps to distinguish the voices (in accordance with Deutsch's observations), but a good violinist will use distinctions in timbre as well to help separate successive notes into two voices, in accord with Wessel's results.

■ A musical illusion described by David Wessel. The notes shown by × and by + in part A may be played with the same or with different timbres. When played with the same timbre (as in part A), they are heard as sequences of three ascending notes. When the timbres differ enough (as in part B), two voices are heard, each consisting of a repeated pattern of three *descending* tones.

## Ciaccona

 A violinist can play at most two simultaneous notes; yet the beginning of this ciaccona by Bach, from his Sonata IV for unaccompanied violin, calls for several four-note chords. Such a chord must be approximated by playing the notes *arpeggio*, in sequence. Furthermore, starting at bar 10, three distinct voices are indicated. The ear can separate these only because the voices move up and down in small intervals and don't overlap in pitch, though any aid that the violinist supplies by differences in loudness or timbre will help.

The foregoing observations indicate that our hearing tends to associate tones of like timbre and to avoid perceived jumps in pitch. A good deal of recent music has made use of widely separated notes in a single melodic line, and some composers have specified that successive notes of a melodic line be played by different instruments. Both practices are hard for performers. In traditional music, large leaps or repeated leaps are used in order to obtain particular effects, such as a dramatic effect in Mozart's great aria for the "Queen of the Night" in *The Magic Flute,* or the effect of a bugle call in *"Non più andrai"* in *The Marriage of Figaro.* Differences of timbre have commonly been used to *distinguish* one musical line from another.

Because of its flexibility and accuracy, the computer can be used to produce auditory illusions that would be difficult or impossible to attain in any other way. In one very striking illusion produced by Risset, the pitch of a sound recorded on tape *falls* slightly when the tape speed is doubled, going from 3.75 inches a second to 7.5 inches a second. Clearly, all sinusoidal components have doubled in frequency. Why has the pitch gone down?

The figure on the facing page shows how this is done. The frequencies of the partials of the tones are plotted as vertical lines on an octave scale, and their intensities are shown as the heights of the lines. The frequencies and intensities of the partials for the tape played at 3.75 inches a second are shown in part A of the figure. All partials are separated by 1.1 octaves. When the tape is played at 7.5 inches a second, the frequency of each partial is doubled and therefore

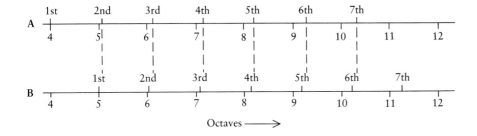

Octaves ⟶

Risset produced an interesting tone in which, if all the frequencies present are doubled, the pitch goes down. The frequencies present are shown in part **A** on an octave scale. The frequency of each partial is 1.1 octaves above that of the preceding partial. When all frequencies are doubled, as shown in part **B**, the listener hears each partial as being replaced by a partial a tenth of an octave lower in frequency, and so hears the pitch as going down. The ear doesn't notice that the old first partial has disappeared and that a new, high-frequency partial has appeared.

shifted up an octave. The new first partial has a frequency about a tenth of an octave below that which the second partial formerly had, the third partial has a frequency about a tenth of an octave below that which the fourth partial had, and so on. In the new sound, most partials are "replaced" by partials about a tenth of an octave lower in frequency, and the ear hears this as a fall in pitch. The fact that the weak first partial has disappeared, and that a new weak partial has appeared an octave above the old seventh partial, passes unnoticed.

Risset's "illusion" is closely related to an earlier illusion devised by the psychologist Roger Shepard, who used a computer to produce a succession of tones that seem to ascend endlessly in pitch by intervals of a semitone. Shepard's illusion is explained in the diagram on the next page. The envelope that specifies the intensity of a partial as a function of frequency is constant, as shown in parts A through E. In the shift from part A to part B, each partial goes up a semitone in frequency; so we hear a change in pitch of one semitone. However, the upper partials have become weaker, and after 12 such shifts, we arrive at the same configuration as shown in part A; so the pitch can continue to change without changing.

One need not proceed semitone by semitone. One can slide slowly up or down the scale, seemingly endlessly. Risset used this very effectively in incidental music for the play *Little Boy*, by Pierre Halet. The theme of the play is the recurrence of the Hiroshima bombing in a nightmare of Eatherly, the pilot of a reconnaissance plane. In this nightmare the bomb falls endlessly in a tone of ever-descending pitch.

Risset has also used tones in which the envelope shifts upward while the partials descend in frequency. The pitch decreases while the tone becomes shriller and shriller.

Kenneth Knowlton of Bell Laboratories and Risset have produced rhythmical sounds that constantly speed up but go slower and slower. As the figure on page 195 shows, the musical beat gets faster and faster, but the sixteenth notes gradually die out in intensity, to be replaced in turn by the eighth notes and the quarter notes.

Risset has produced subtle pitch paradoxes, in which the pitch of a bell-like sound with nonharmonic partials is easily identified with that of either of its two most intense sinusoidal partials.

■ Shepard's famous eternally ascending tones. These tones are made up of octave partials whose amplitudes are specified by the fixed envelope shown in parts **A** through **E.** In the shift from part **A** to part **B,** each partial is increased in frequency by a semitone, and we hear a rise in pitch of a semitone. Similarly for the shift from part **B** to part **C.** After 12 increases by a semitone, we will be back at the same configuration as in part **A;** so, if we continue, we keep hearing a never-ending sequence of increases in pitch.

Octaves ——→

2 3 4 5 6 7 8

A

B

C

D

E

■ Maurits Escher's *Ascending and Descending.*

*accelerando*

■ By analogy with the process illustrated in the preceding figure, Risset has produced a rhythm that goes faster and faster and yet slows down. Although the beat gets faster and faster, the shorter notes are gradually replaced by longer notes.

Illusions of moving sounds are among the most effective of computer-produced effects. Much of the background has been covered in earlier chapters. Sound sources can be made to stand out, filling a room. Individual sources can whirl about over one's head as in John Chowning's *Turenas*. Such effects are most striking with four-track recording and four loudspeakers at the four corners of a square, as Chowning demonstrated. The position of the listener relative to the loudspeakers may be less important for them than for reproduction of instrumental sounds by a quadraphonic system.

The principal effects to be produced are: (1) direction (azimuth) of the sound source; (2) distance of the sound source, and (3) motion of the sound source.

Chowning's prescription for direction is to divide the intensity of the sound source between two loudspeakers at the left and right of the source, using signals of the same phase or delay.

Chowning's prescription for distance is to control the ratio of reverberant to direct sound. Nearby sources will have more direct than reverberant sound; distant sources will have more reverberant than direct sound. The amplitude of the direct sound should change as the reciprocal of the distance (the intensity as the reciprocal of the square of the distance). In a small room the intensity of the reverberant sound changes little with distance; in a large room it decreases somewhat with increasing distance of the source. Chowning has decreased the intensity of reverberant sound with the reciprocal of the distance of the source.

Slow source motions can be simulated simply by changing the direction and distance of the source. For swifter motion, the Doppler effect can be incorporated. The frequencies of sound sources moving toward us are raised; those of receding sources are lowered. A classical example is the sound of the whistle of a locomotive as it approaches, passes, and recedes. If the sound source approaches at a speed $s$, all frequencies are raised by a fraction $s/v$, in which $v$ is the velocity of sound. If the source recedes at a speed $s$, all frequencies are lowered by a fraction $s/v$. To change the pitch by one semitone, the speed relative to the listener must be about 0.06 that of sound, that is, about 21 meters per second, or 68 feet per second.

In producing distance cues, some sort of artificial reverberation is necessary. In effect, fractions of the original computer-generated sound are added in with various delays. A great deal of study has been devoted to getting natural-sounding reverberation that does not alter the perceived spectrum of sounds. This can be done only if the reverberant signal has the same power spectrum as the original signal (colorless reverberation, first described by Manfred Schroeder) or if any alterations in spectrum change very rapidly with frequency, so that the reverberant intensity averaged over a critical bandwidth is little changed.

Although the best illusions of moving sound sources are attained with quadraphonic systems, startling effects are possible with only two loudspeakers. Not only do sound sources seem to move about; they stand out so that the sound fills the room in a way that seems unrelated to the positions of the loudspeakers.

Twice I have experienced the effect of sounds coming from unexpected directions. In the home of a friend I was listening to a recording of some rather ordinary music played on a synthesizer. The music was embellished with high-frequency chirping sounds. Some of these seemed to come from the left or right of the stereo loudspeakers, and even from the back of the room. On another occasion I was listening to a stereo radio broadcast that included high-pitched bird sounds. Some of these came from most unexpected directions, and certainly far to the right or left of the loudspeakers. What can the explanation be?

Perhaps, in some sense, the sounds from the loudspeakers happened to mimic Schroeder's perfect stereo system (the one using a dummy head to pick up sounds). Perhaps the explanation is simpler. The effect may have been caused by the directional patterns of the loudspeakers at very high frequencies, coupled with reflections from ceiling and walls. Perhaps both explanations are the same. What I *do* know is that a two-channel stereo system can cause high-frequency sounds to be heard as coming from almost any direction. I don't see why this effect could not be exploited using computer-generated sounds.

Indeed, many strange effects are possible. At Stanford I heard a sound generated by Malcolm Singer that, starting from a single loudspeaker a couple of yards away, rushed out until it enveloped my head. The pitch began low and ended high. I won't try to explain the mechanism of what I heard. But, as noted early in Chapter 7, the pinnae of our ears are essential in judging the height of sound sources, and, presumably, in telling whether the source is in front of us or behind us. They can do this only by modifying the high-frequency spectra of the sounds that reach them, and our judgment of up or down, ahead or behind, must be based on such modifications. Surely it must be possible to modify the high-frequency spectrum of a sound deliberately, to give a sense of the sound's moving up, down, backward, or forward.

Whatever effects we may be able to attain by using computer-generated sounds, we want them to be of good quality. Early computer-generated sounds were buzzy or "electronic." Because our understanding of the nature of good instrumental sounds has been improved by computer analysis and synthesis of sounds, we now have computer-produced sounds that are neither harsh nor electronic. Some are indistinguishable from sounds of instruments that they imitate. Some sound like nothing anyone has ever heard before, for they involve carefully controlled nonharmonic partials, or strange variations of partials that could not be produced by conventional instruments or, in fact, any imaginable mechanical instruments. In some computer-produced sounds, one sound is transformed into another: a bell into prolonged liquid textures or into a group of singing voices, a voice into the roar of a lion. In others, "nonexistent" sound sources move through unoccupied space: ghostly instruments glide through an empty room.

People have learned to attain such effects because of their exploration of sounds and their efforts to characterize musical sounds. One great puzzle in the characterization of musical sounds is the issue of *categorical perception,* which is characteristic of our perception of speech sounds called *phonemes.* These form the alphabet of spoken language. Each language has a specific number of phonemes. The common consonant and vowel sounds of English are the phonemes of the English language. Phonemes differ from language to language.

The phoneme is the *percept,* not the physical sound. In different words the sound wave by which we recognize *b, g, k, o,* or *u* can differ. In common speech, the range of difference is small enough that we can recognize the phoneme correctly, whatever its context. We seldom confuse *got* with *cot,* even though the *g* (voiced) sound is very like the *k* (unvoiced) sound. Nor do we confuse *had* with *hod.*

By producing speech sounds artificially, we can make a gradual transition between the sound wave characteristic of one phoneme and that characteristic of another. When we do this, the listener hears *either* one phoneme *or* the other, not something in between. This is called categorical perception.

Categorical perception is merely categorical; it is not necessarily accurate. The hearer may hear a boundary-line sound sometimes as *g*, sometimes as *k*, but never something "in between." An English speaker has never learned anything in between, but only the limited number of phonemes characteristic of our language.

We are led to ask, is there categorical perception of musical sounds? It is easy to jump to the conclusion that there is. An expert musician has no difficulty in saying, "That's a violin." Or a viola, or a French horn, or a saxophone. That sounds like categorical perception. But when John Grey synthesized musical tones midway between the cubes representing two instruments in the three-dimensional space shown in the figure on page 179, these tones weren't recognized as one instrument or the other. Rather, sounds seemed related to the two instruments, or a mixture of them. This is contrary to the categorical perception of a speech sound as one phoneme *or* the other.

In certain respects we may experience something like categorical perception in listening to musical sounds, especially in our experience of pitch. The clearest examples are mistakes of an octave. A musician may mistake the pitch of some sounds by an octave, but he makes a mistake of exactly one octave. It is somewhat similar with the notes of the scale. The expert listener correctly identifies the notes of the justly tempered scale, the Pythagorean scale, or the equal-tempered scale, despite differences in frequency. When a singer sings a little off pitch, the expert listener characterizes the note as flat, not as a different intended pitch. This is a little like recognizing a phoneme and a foreign accent at the same time. Beyond this, expert listeners recognize chords in a categorical manner, regardless of what instruments play them.

In conventional western music, only certain pitches are intended (or "allowed"), just as in English only certain speech sounds are intended (or "allowed"). It therefore would not be surprising to find categorical perception among pitches, and among chords as well. In Chapter 6, I noted that musicians correctly recognize the dominant seventh chord, even when it has been doctored into being consonant.

Categorical perception of timbres is another matter; yet I believe there is something of the sort, though it is weaker than with speech. We won't find this sort of categorical perception by wandering through the three-dimensional timbre space of the figure on page 179. That space is too limited. If there are musical categories of sound, they are not trumpetlike or trombonelike, but reedlike, brassy, bowed, struck or plucked, bell-like or gonglike (nonharmonic), drum-like, blocklike. We certainly recognize natural (noncomputer) sounds as belonging to various categories, which are associated with the nature of the sound-producing material and with its mode of excitation.

Would categorical perception of musical sounds be a help or a hindrance? The scale and common chords give coherence to music. But isn't it tempting to think of *new* timbres that will *sound* new and different? And even of new scales and chords? We have learned language so thoroughly that we make only categorical distinctions between a relatively few speech sounds, hearing nothing in between. It is hard or impossible for an adult to undo this training. Adult Japanese find it very difficult to learn to hear the English *r* and *l* as different, though Japanese children learn readily. It may be difficult for us to hear, to distinguish, to recognize new and unfamiliar musical sounds. Happily, it seems not to be impossible.

In the nineteenth century Helmholtz performed miracles with very simple equipment in the analysis and understanding of musical sounds. In the first half of this century, the electronic art derived from telephony made possible more quantitative and subtler experiments. In our own time the computer has made easy what was difficult or impossible with earlier electronic means. It has done something more. By using the computer we can transcend all the limitations of early sound sources. We can imitate the sounds of fine instruments. We can go beyond them. Our understanding of musical effects has increased, and so has our ability to produce them. What can come of this?

As I noted in the first chapter of this book, an increased capability to generate, experiment with, and understand sounds, new and old, has led some enterprising composers to pay more attention to the subtle qualities of the sounds used in their compositions. This seems to me to be a healthy alternative to excessive concern with formal structure, or to a search for "spontaneity" based on some form of improvisation.

Both rules and spontaneity have their place in music, but deep understanding and careful work are essential, too. However much we may wish that we could have heard the improvisations of Bach, Mozart, or Debussy, it seems likely that their best music is the music that they left us.

The genius of the past can be a weight upon the present. How shall music avoid being crushed by it? Perhaps by using new resources and new understanding. Scientists are not crushed by Newton and Einstein, for they have experimental resources, knowledge, and understanding that Newton and Einstein lacked. Whatever its "absolute" value, new science is new and worthy when it succeeds in going successfully beyond the old.

May it not be so with new music? But to succeed, new music must really be heard in the sense that the composer intended, must be understood, must hold the interest of and move the listener. Here an understanding and exploration of the science of musical sounds can help. The rest only talent or genius can supply.

# Appendixes

## Appendix A  Terminology

When scientists and engineers deal with well-defined, measureable physical properties, they use only unambiguous, well-defined terms, such as *time,* measured with a watch, *mass,* measured with a balance, or *length,* measured with a meter stick. This helps one grasp their intended meaning.

Many legitimate physical terms aren't nearly so simple. They can't be explained in a few words. Understanding them and using them properly results only from long exposure to experience or experiment, until the terms and their place in physics become familiar and, indeed, commonplace.

Efforts to "define" words briefly in terms of other words aren't very helpful. In both everyday life and science (if not in philosophy), we learn to use words understandably by protracted experience with things and by communicating with others. In this book I have tried to use words understandably, but I finally had to give up trying to eliminate every vestige of ambiguity.

In music we deal with many difficult qualities. They would not be made less difficult if I were to depart from the common words used by musicians, to invent a jargon, or to import one from psychology. I believe that the chief difficulties lie in the facts and experiences themselves, not in the words we use about them. I believe that the best "definitions" of the words that I use are in the text, explicitly or implicitly. Nonetheless, this brief discussion of terminology may prove useful to the reader.

Strictly, a *sound* is what we hear when a *sound wave* going through the air strikes our ears. A sound wave acts as what psychologists call a *stimulus.* Our *response* to the stimulus is the "sound" that we hear. By this definition, if no one is listening (or if only deaf persons are present), then there is no sound, but only a sound wave in the air.

The word *note* can designate *either* a mark on a musical staff *or* the sound produced, which we hear when someone "plays a note." Some try to avoid confusion by using the word *tone* for the sound produced when someone "plays a note."

A tone is a *musical* sound, one that can be heard as having a *pitch*. One can apply "tone" to the sound of a bell, but not to the sound of a drum. Musical sound waves are periodic fluctuations of air pressure. A *pure tone* is a sinusoidal sound wave. (One can also speak of the *good tone* of a violin, violinist, or pianist, but in this book I try not to use *tone* in that sense.)

*Pitch* is a quality that we hear in some sounds. Happily, for real periodic sounds the pitch that we hear is tied firmly to the periodicity or frequency of the sound wave. In concert pitch, A above middle C has a frequency of 440 Hz (vibrations per second). I therefore think it proper to designate pitch quantitatively by specifying frequency.

*Loudness* is how loud a sound sounds. It is related in a complicated way to the *intensity* of a sound. Intensity is measured in watts per square meter, a good, solid physical quantity.

*Timbre* is a quality that a sound has in addition to pitch and loudness. Sounds that don't have a clear pitch, such as those of drums and blocks, can differ in timbre. We can use many common words to distinguish timbres: *shrill, warm, harsh, dull, percussive.* Such words describe real, consistent differences in our responses to musical sounds and in the sound waves, but it is no simple matter to pin these differences down.

Is it differences in timbre that distinguish good from bad violin playing? Physicists rightly assert that a piano emits the same sound whether an expert player strikes a key or a weight falls on it; yet some pianists obviously have a "good tone" and others don't. I don't know how to explain this difference, and you won't necessarily get help by asking a pianist who has a "good tone." She (or he) can produce the effect, but will probably not be able to tell you in words how this is done. "Make it sing," Claude Shannon's clarinet teacher told him. Shannon knew what was wanted, if not how to do it.

# Appendix B  Mathematical Notation

I have tried either to avoid mathematics in the text or to make it as simple as possible. Some mathematical expressions are necessary in conveying quantitative relations. For example,

$$ml$$

means $m$ times $l$. Clearly, this won't work for numbers, for we would not know whether

$$27$$

was two times seven or twenty-seven. Hence, when we want to indicate the multiplication of numbers, we enclose them in parentheses. Thus

$$(2)(7)$$

is 2 multiplied by 7. We can do this, if we wish, in multiplying quantities represented by letters,

$$(m)(l) = ml,$$

but nothing is gained by including the parentheses.

The expression

$$t^3$$

is the *third power* of $t$ (or the *cube* of $t$). The meaning is

$$t^3 = ttt = (t)(t)(t)$$

The *exponent*, 3, tells how many $t$'s to multiply together. A negative exponent indicates division rather than multiplication. Thus,

$$t^{-3} = 1/t^3 = (1/t)(1/t)(1/t).$$

Let's consider a numerical example of powers. In an equal-tempered scale the frequency ratio of a semitone is

$$(2)^{1/12} = 1.059468.$$

Sometimes a very large number is written in the following way:

$$5.4 \times 10^5 = (5.4)(10)(10)(10)(10)(10)$$
$$= 540,000;$$

We could write $5.4 \times 10^5$ as

$$(5.4)(10^5)$$

but for some reason we don't. A very small number can likewise be written as

$$6.2 \times 10^{-4} = 6.2(1/10)(1/10)(1/10)(1/10)$$
$$= 0.00062.$$

As a review,

$$ml/t^2$$

is the product of $m$ and $l$ divided by $t^2$.

The *square root* of a quantity, say, $x$, is such that

$$(\sqrt{x})(\sqrt{x}) = x.$$

Thus,

$$\sqrt{4} = 2$$

and

$$(\sqrt{4})(\sqrt{4}) = (2)(2) = 4.$$

We encounter decibels, abbreviated dB, in several chapters. Decibels provide a way of expressing ratios of powers. If $P_1$ and $P_2$ are two powers (commonly measured in watts), $P_2$ is greater than $P_1$ by

$$10 \log_{10} (P_2/P_1) \text{ dB}$$

The logarithm to the base 10 of a number can be looked up in a table or obtained by using a "mathematical" hand calculator. The table at the left gives the flavor of logarithms.

Logarithms and decibels.

| Power ratio $R$ | Amplitude ratio ($\sqrt{R}$) | $10 \log_{10} R$ |
|---|---|---|
| .0001 | .01 | $-40$ dB |
| .001 | .0316 | $-30$ dB |
| .01 | .1 | $-20$ dB |
| .1 | .316 | $-10$ dB |
| 1 | 1 | 0 dB |
| 10 | 3.16 | 10 dB |
| 100 | 10 | 20 dB |
| 1,000 | 31.6 | 30 dB |
| 10,000 | 100 | 40 dB |
| 2 | 1.4 | 3 dB |
| 1/2 | .71 | $-3$ dB |

# Appendix C  Physical Quantities and Units

The MKS or meter-kilogram-second system of units is used. The units in which quantities are measured are:

| | |
|---|---|
| mass, $m$ | kilogram |
| distance, $l$ | meters |
| time, $t$ | seconds |
| force, $F$ | newtons |
| power, $P$ | watts |
| intensity, $I$ | watts per square meter |
| energy, $E$ | joules |

The pull (or "acceleration") of gravity at the Earth's surface, which is 9.80 meters per second per second, produces a force of 9.80 newtons on a mass of one kilogram. Force is mass times acceleration.

A force of one newton pushing something a distance of one meter requires an expenditure of one joule of energy.

The number of watts is the number of joules of energy expended per second.

In MKS units, pressure is measured in newtons per square meter. One newton per square meter is called a *pascal*.

# Appendix D  Mathematics and Waves

Chapter 2 correctly describes the propagation of waves along a string or through the air as a traveling disturbance. Such a disturbance involves continual changes in momentum (mass times velocity) caused by a force. The force may be associated with the bending of a stretched string or the compression of air. Continual changes in the force occur because the stretched string is bent as the wave travels along, or because the air is compressed when the velocity associated with the wave is lower ahead than it is behind.

The propagation and properties of both transverse and longitudinal waves can be demonstrated by simple but rather tedious mathematical analysis (reasoning). One result of such mathematics is the fact that waves behave in a very simple fashion only for small amplitudes; that is, when the stretched string along which a wave travels isn't bent too sharply or when a sound wave traveling through the air raises or lowers the pressure only by a small fraction. The behavior of such small-amplitude waves is *linear*. In essence, this means that when two waves are present in the same medium (string, air), they don't interact with one another. Each goes its own way as if the other weren't present. The total motion (displacement, velocity, or pressure) is simply the sum of the motions associated with the two (or more) waves.

We will not attempt a conventional mathematical analysis of waves here. Instead, we will assume that we are dealing only with small-amplitude linear waves. We will then ask, how can the velocity of the waves, the power carried by them, and other properties be expressed in terms of various physical properties? We do this by a seeming magic called *dimensional analysis*.

All physical quantities—including force, velocity, and momentum—have a dimension that is expressed in terms of the dimensions of time, mass, and length. Actual time measured in seconds is designated herein by the italic letter $t$, whereas the dimension of time is designated by the script letter $t$, and similarly for mass, length, and any other physical quantity and its dimension. The three fundamental physical quantities and their dimensions are given in the table below.

| Physical quantity | Symbol for physical quantity | Symbol for dimension of physical quantity |
|---|---|---|
| Time | $t$ | $t$ |
| Mass | $m$ | $m$ |
| Length | $l$ | $l$ |

Let us illustrate the dimensions of some common physical quantities. What is the dimension $f$ of frequency or periodicity in time? Frequency is the number of something per second. Number is dimensionless, and so the dimension of frequency is simply

$$f = 1/t. \tag{D-1}$$

What about velocity? Velocity is distance per unit time. The dimension of distance is $l$, and the dimension of time is $t$; so $v$, the dimension of velocity, is given by

$$v = l/t. \tag{D-2}$$

Acceleration is change in velocity with time. Numerically, it is the amount that velocity changes in a unit of time. Dimensionally, the dimension $a$ of acceleration is given by

$$a = v/t = l/t^2. \tag{D-3}$$

We now come to a physical law, not a matter of definition. This law was first stated by Newton. It is that, numerically, force is equal to mass times acceleration. Thus, the dimension $\mathcal{F}$ of force is

$$\mathcal{F} = ml/t^2. \tag{D-4}$$

Energy (or work) can be defined as force times distance. Thus, the dimension of energy, $E$, is

$$\mathcal{E} = ml^2/t^2. \tag{D-5}$$

Let us take a look at equation D-5. We note from equation D-2 that the dimension of velocity is $l/t$. Thus, the dimension of energy can also be written

$$E = mv^2. \qquad (D-6)$$

Dimensionally, this is correct. But the numerically correct expression for kinetic energy, or the energy of mass in motion, is

$$E = (1/2)mv^2. \qquad (D-7)$$

Here $E$ is actual energy, not the dimension of energy, $m$ is actual mass, and $v$ is actual velocity. Dimensionally, equations D-6 and D-7 are in accord. The numerical factor (1/2) is a number, and has no dimension. What we *can* say is that finding the dimension of energy by dimensional analysis has led us to an expression for kinetic energy that is correct in everything except a multiplying numerical factor.

Let us keep what we have learned in mind, and see how we can use dimensional analysis in connection with waves.

As an example, what is the expression for the velocity $v$ of a transverse wave traveling along a string of mass $M$ kilograms per meter, a string that is stretched with a tension or force of $T$ newtons?

First, what are the dimensions of $M$ and $T$? $M$ is mass per unit length; so the dimension of $M$ is

$$M = m/l. \qquad (D-8)$$

The tension $T$ is simply a force, which has dimensions $ml/t^2$, so the dimension of $T$ is

$$T = ml/t^2. \qquad (D-9)$$

The dimension of the velocity $v$ of the wave must be $l/t$. We can see that this will be true if the expression for the velocity is

$$v = \sqrt{T/M}. \qquad (D-10)$$

We verify this by writing for the dimensions of $\sqrt{T/M}$,

$$\sqrt{T/M} = \sqrt{(ml/t^2)/(m/l)}$$
$$= \sqrt{l^2/t^2}$$
$$= l/t = v. \qquad (D-11)$$

Actually, equation D-10 is the numerically cor-

rect expression for the velocity of a wave traveling along a stretched string. The numerical factor turns out to be unity, but we couldn't know this from dimensional analysis.

Let us turn to something that is very important about plane waves traveling through air, that is, the intensity $I$, which is the power density measured in watts per square meter.

We noted earlier that energy or work can be defined as force times distance. Power is energy per unit time; so power will have the dimensions of force times ($l/t$), or force times velocity. Intensity $I$ is power per square meter; so intensity will have the dimensions of force times velocity divided by $l^2$. Hence, the dimension $I$ of intensity is given by

$$I = F(l/t)/l^2 = F/lt = m/t^3. \qquad (D-12)$$

But in plane waves in air we are concerned not with force $F$, but with force per square meter, or pressure, $p$. The dimension $p$ of pressure is

$$p = F/l^2. \qquad (D-13)$$

From these last two relations we see that the dimension $I$ of intensity is

$$I = p(l/t). \qquad (D-14)$$

Here $p$ is the dimension of pressure, and $l/t$ is the dimension of velocity. We may easily conclude that if $p$ is the fluctuating pressure associated with a sound wave and $u$ is the fluctuating velocity associated with a sound wave, the intensity of the wave will be

$$I = pu. \qquad (D-15)$$

This is not only dimensionally correct, but also numerically correct, and we could have arrived at it more directly.

In a small-amplitude linear sound wave, the pressure $p$ will be some constant, which we will call $K$, times the velocity fluctuation $u$:

$$p = Ku. \qquad (D-16)$$

Thus, by using equation D-16, we can express intensity $I$ in terms of either $p$ or $u$:

$$I = Ku^2, \qquad (D-17)$$
$$I = (1/K)p^2. \qquad (D-18)$$

$K$ is called the *characteristic impedance* or *wave impedance* for a plane sound wave. But how are we to find an expression for $K$? The dimension of $K$ is

$$K = I/v^2 = I(l/t)^2. \qquad (D-19)$$

Hence, from equations D-19 and D-12,

$$K = m/l^2t = (m/l^3)(l/t). \qquad (D-20)$$

The first factor on the far right of equation D-19 has the dimensions of mass density, which we will call $D$. The second factor has the dimensions of velocity. Can it be that

$$K = Dv, \qquad (D-21)$$

in which $D$ is the density of the air and $v$ is the velocity of sound? It can be and it is, and equation D-21 is numerically correct. This is very plausible, for from equations D-21 and D-17 we can write

$$I = (Du^2)v. \qquad (D-22)$$

$Du^2$ is proportional to the kinetic energy per cubic meter of the air moving at a velocity $u$, and in some sense this energy is transported through the air at a rate $v$. The kinetic energy is only half the energy transported; there is an equal amount of potential energy associated with the compression of the air by the sound wave.

We can also express the intensity of the sound wave in terms of the pressure $p$ as

$$I = p^2/Dv. \qquad (D-23)$$

For air at 20° C,

$$D = 1.2174 \text{ kilograms per square meter,}$$
$$v = 344 \text{ meters per second,} \qquad (D-24)$$
$$I = 0.002388\ p^2.$$

We should note that for a fluctuating pressure, the average intensity is given by the average value of $p^2$ divided by $Dv$. For a sinusoidal variation of pressure with time, the average value of $p^2$ is half the square of the peak pressure (that is, the "top" of the sine wave).

In Chapter 7 the reference level of intensity is given as $10^{-12}$ watt per square meter. The usual reference level is a pressure of 0.00002 pascal (a pascal is a pressure of one newton per square meter). If we calculate $I$ for this pressure using (D24), we get

$I = 0.955 \times 10^{-12}$ watt per square meter.

This is so close to the $10^{-12}$ watt per square meter (only 0.2 dB different) that I chose to use the round number $10^{-12}$ watt per square meter in Chapter 8.

In Chapter 8 I commented on the sensitivity of the ear, and said that "theoretically" we should just be able to hear a 1-watt, 3,500-Hz sound source at a distance of 564 kilometers (352 miles). If a sound has a power of $W$ watts and travels out equally in all directions, so that at a distance $L$ the power passes evenly through a sphere of area $4\pi L^2$, the intensity $I$ at a distance $L$ must be

$$I = W/4\pi L^2. \qquad (D\text{-}25)$$

If we let $W = 1$ and $L = 564,000$ meters, we get from equation D-22 just about $10^{-12}$ watt per square meter, which is about the threshold of hearing.

Let us turn to the velocity with which a sound wave travels through air. This velocity does not vary with pressure, but it does vary with temperature. The pressure of air is caused by the velocity of air molecules. The square of this velocity is proportional to the temperature in kelvins, that is, in degrees measured with respect to absolute zero; that is, 0 kelvin is $-273$ degrees Celsius (centigrade).

It is dimensionally plausible that the velocity of a sound wave should be proportional to the velocity of the molecules of the air through which the wave travels, and this turns out to be so. The velocity of a sound wave at a temperature $T$ can thus be expressed as

$$v = v_k \sqrt{T/T_k}, \qquad (D\text{-}26)$$

in which $v_k$ is the velocity of the sound wave at a temperature of $T_k$ kelvins. If we take the velocity of the sound wave to be 344 meters per second at a temperature of 20 degrees Celsius (centigrade), then, at a temperature $T$ kelvins,

$$v = 344 \sqrt{T/293},$$
$$v = 20.1 \sqrt{T}. \qquad (D\text{-}27)$$

The velocity of sound varies with humidity as well as with temperature. The velocity of the molecules of a gas varies with the mass of the molecules as well as with the temperature: the lighter the molecules, the greater their velocity. Molecules of water vapor are less massive than molecules of dry air, and so the velocity of sound increases with increasing humidity.

Wind instruments are provided with tuning adjustments to compensate for the effects of temperature and humidity on pitch. The pitch of a pipe organ inexorably changes with temperature and humidity, and other instruments must accomodate this.

## Appendix E  Reflection of Waves

In order to understand the reflection of sound waves, we must take into account the fact that such waves consist both of an increase or decrease in the pressure of the air, designated by $p$, and of a forward or backward velocity of the air, designated by $u$. As noted in Appendix D, these two components of the wave go hand in hand. In a wave that travels from left to right,

$$p = Ku,$$

in which $K$ is a constant.

The figure below shows graphically the pressure $p$ and the velocity $u$ of a "square" wave traveling to the right. Part A depicts a wave in which the pressure $p$ is positive. The associated velocity $u$ is positive; that is, it represents motion of the air to the right, in the direction in which the wave travels. In part B, the pressure $p$ is negative, and is represented as lying below the axis, the horizontal line marked 0. This merely means that the total pressure of the air is less than the average air pressure. The velocity $u$ is also negative (motion of the air to the left), and this is shown as a velocity lying below the axis.

The figure at the bottom of the facing page shows $p$ and $u$ for a square wave traveling to the left. Why do the graphs of waves traveling to the right look so different from those of waves traveling to the left? A wave can travel to either right or left. Shouldn't the picture be much the same in either case? The difference arises because in both cases we measure a velocity as positive (above the horizontal axis) if it is directed toward the right, regardless of whether the wave travels to the left or to the right. In any sound wave, if the pressure $p$ is positive, the velocity of the air, $u$, is in the direction in which the wave travels, and if the pressure $p$ is negative the velocity of the air is directed opposite to the direction of travel of the wave. Thus, in the graphs of a wave traveling to the left, the pressure $p$ is positive and the velocity $u$ is negative, that is, to the left, the direction in which the wave travels.

Graphs such as those below and those at the bottom of the facing page are very useful for understanding the reflection of waves. The figure at the extreme right on the facing page illustrates successive stages in the reflection of a wave from a wall, represented by the vertical

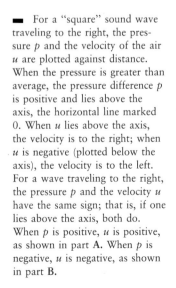

■ For a "square" sound wave traveling to the right, the pressure $p$ and the velocity of the air $u$ are plotted against distance. When the pressure is greater than average, the pressure difference $p$ is positive and lies above the axis, the horizontal line marked 0. When $u$ lies above the axis, the velocity is to the right; when $u$ is negative (plotted below the axis), the velocity is to the left. For a wave traveling to the right, the pressure $p$ and the velocity $u$ have the same sign; that is, if one lies above the axis, both do. When $p$ is positive, $u$ is positive, as shown in part A. When $p$ is negative, $u$ is negative, as shown in part B.

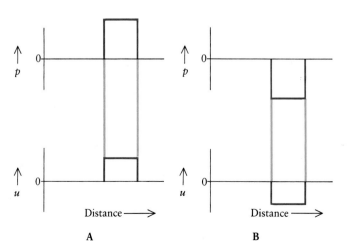

A                                   B

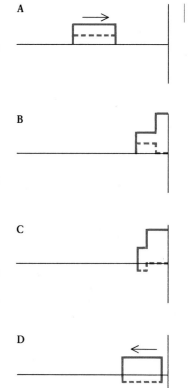

line at the right. In this figure, pressure *p* is represented by a solid line and velocity *u* by a dashed line. In part A, we see the pressure and velocity of an *incident* wave approaching from the left. Both are positive, as is proper for a wave traveling to the right. In part D, the wave has been *reflected* from the wall. The pressure is positive and the velocity is negative, as is proper for a wave traveling to the left.

What of the pressure and velocity during the process of reflection, as shown in parts B and C? Here the pressure near the wall is twice as great as that of the incoming wave, and the velocity near the wall is zero. Can we explain this complex behavior?

The behavior is actually very simple. During the process of reflection we have overlapping waves, one traveling to the right and the other to the left. The pressure and velocities shown in parts C and D are the sums of the pressures and velocities of these two waves: the pressures add, but the velocities, being in opposite directions and therefore having opposite signs, cancel each other out.

■ For a wave traveling to the left, the pressure *p* and the velocity *u* have opposite signs. When the pressure is positive (as in part A), the velocity is negative. When the pressure is negative (as in part B), the velocity is positive. This is so because a positive velocity means a velocity to the right. Thus in part A the pressure is positive, and the velocity is in the same direction as that in which the wave travels, that is, to the left, shown as a negative velocity.

■ The reflection of a sound wave from a solid wall, represented by the vertical line to the right. Pressure is represented by a solid line and velocity by a dashed line. Part A represents a wave traveling to the right, toward the wall; part D, a wave traveling to the left, away from the wall; parts B and C, the wave in the process of reflection. The complicated curves of pressure and velocity in parts B and C are simply combinations of the pressures and velocities of waves traveling to the right (the *incident* wave) and to the left (the *reflected* wave). The velocities of these two waves must be such that at the wall the sum of the velocities is zero, for the air can't move at the wall. This simple condition determines the velocity, and hence the pressure, of the reflected wave.

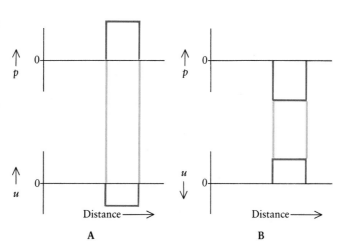

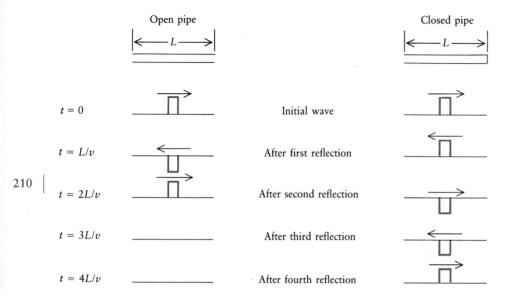

| | Open pipe | | Closed pipe |
|---|---|---|---|
| | Open pipe | | Closed pipe |

$t = 0$ — Initial wave

$t = L/v$ — After first reflection

$t = 2L/v$ — After second reflection

$t = 3L/v$ — After third reflection

$t = 4L/v$ — After fourth reflection

We see that, in reflection from a solid obstacle, the pressure of a sound wave is reversed. When a sound wave travels through a tube, we can have another sort of reflection. If the end of the tube is open, there can be no (or very little) pressure at the end of the tube, but we can have a velocity $u$ at the end of the tube.

Imagine a sound wave traveling through a narrow tube, such as an organ pipe. When it reaches an open end, it will be almost completely reflected. But, after reflection at an open end, the air velocity $u$ will be unchanged and the sign of the pressure $p$ will be reversed. If the pressure $p$ is positive before reflection, it will be negative after reflection.

In relating the length of a pipe to pitch, we must take into account the nature of the reflections at its ends. The figure above illustrates successive reflections in an organ pipe open at both ends (left) and in a pipe open at one end and closed at the other (right).

In reflection from an open end, the pressure $p$ changes sign on reflection, but, in reflection from a closed end, the sign of the pressure $p$ remains the same after reflection. For an open pipe, we see that after a time,

$2L/v,$

the wave has undergone two reflections and is the same as at the start. However, for a closed pipe the wave must undergo four reflections be-

fore it is the same as at the beginning, and this takes a time

$4L/v.$

Thus the pitch frequency $f$ of an open organ pipe is

$$f = v/2L \text{ Hz}, \qquad (\text{E-1})$$

whereas the pitch frequency $f$ of a pipe closed at one end and open at the other is

$$f = v/4L \text{ Hz}. \qquad (\text{E-2})$$

Here $v$ is the velocity of sound and $L$ is the length of the pipe.

In stringed musical instruments, both ends of the string are rendered immovable. Hence, at the ends of the string, the transverse velocity is always zero. All reflections are the same, and the pitch frequency is always

$$f = v/2L \text{ Hz}. \qquad (\text{E-3})$$

But, in this case, $v$ is the velocity with which a transverse wave travels along the string. As noted in Appendix D, this velocity is higher the greater the tension, and lower the greater the mass of the string.

■ The pressure of a wave after successive reflections at the ends of an open organ pipe (open at both ends, left) and a closed organ pipe (open at one end, closed at the other, right). In the open pipe, after two reflections the pressure is back where it started, and the pitch frequency is $v/2L$ Hz. In the closed pipe, the pressure is back where it started only after four reflections, and the pitch frequency is $v/4L$ Hz. For the same pitch, open organ pipes must be longer than closed organ pipes.

# Computer Generation of Sound

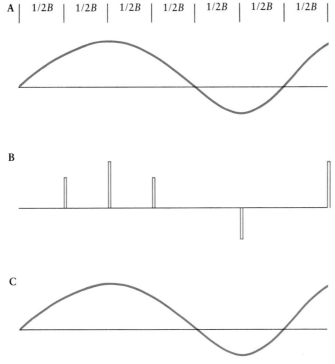

A | 1/2B | 1/2B | 1/2B | 1/2B | 1/2B | 1/2B | 1/2B |

B

C

■ A waveform of bandwidth $B$ can be represented *exactly* by $2B$ samples a second, taken at time intervals $(1/2)B$. In part **A**, the samples are represented by the lengths of vertical lines drawn from the horizontal line to the curve representing the waveform. The sample amplitudes may be described by numbers. In part **B**, the sample amplitudes are represented by very short electric pulses whose heights correspond to the sample amplitudes. In part **C**, the original waveform is recovered by low-pass filtering of the sequence of pulses shown in part **B**.

Anyone who wants to understand in detail the computer generation of sound, or to undertake to generate such sounds, should read Max V. Mathews's book, *The Technology of Computer Music,* which discusses fundamental principles and explains Music V, the last of five programs that Mathews wrote for the computer generation of sounds. Other programs, such as the Music 11 and Music 360 of Vercoe, are adaptations or extensions of an earlier program of Mathews, Music IV, and the ideas of Mathews's music programs permeate all computer generation of sound.

How is it possible for a computer to generate sounds? The *sampling theorem* gives us a clue. Consider any waveform made up of frequency components whose frequencies are less than $B$. That is, consider any sound wave whose frequency components lie in the bandwidth between zero and $B$. *Any* such waveform or sound wave can be represented *exactly* by the amplitudes of $2B$ *samples* per second. These samples are merely the amplitudes of the waveform at sampling times spaced $1/2B$ apart in time. Part A of the figure at the left represents a waveform. In part B, the amplitudes at times $1/2B$ are shown as vertical bars drawn from the horizontal axis (which represents zero amplitude) to the curve of these bars. The successive heights can be represented by $2B$ *numbers* each second. These numbers describe the required samples from which the waveform can be reconstructed. Part C is an exact replica of the waveform in part A. This replica can be obtained by passing short electric pulses of the heights given in part B and described by the numbers that express these heights, through a low-pass filter of bandwidth $B$.

In high-quality computer-generated sound, it is customary to use 50,000 samples per second to represent a waveform. This allows us to produce frequencies up to 25,000 Hz. Because of technological limitations, the frequency range or bandwidth actually attained is 20,000 Hz or less.

The samples in part B of the figure on the preceding page are shown as lines of various heights, and the succession of sample heights stands for a succession of numbers. A computer can't produce every exact number, for most numbers can be represented only by an infinite number of digits to the right of the decimal point. A computer can produce a set of numbers such as 00 (0), 01 (1), 05 (5), 27, 44, 99—all of which are examples of the 100 possible two-digit numbers starting with 00 and ending with 99.

The internal organization of computers is such that they use only two *binary digits,* 0 and 1, and represent numbers in terms of them.

The successive digits of common or *decimal* numbers are interpreted by means of powers of 10. Thus, 257 means

$$7 \times 10^0 + 5 \times 10^1 + 7 \times 10^2$$
$$= 7 \times 1 + 5 \times 10 + 7 \times 100.$$

The binary number 1001 means

$$1 \times 2^0 + 0 \times 2^1 + 0 \times 2^2 + 1 \times 2^3$$
$$= 1 \times 1 + 0 \times 2 + 0 \times 4 + 1 \times 8.$$

In decimal notation, the binary number 1001 is 9.

In good computer-produced sounds, 16 binary digits are used to represent the samples. This enables us to represent 65,536 different sample amplitudes. In effect, half of these binary numbers are used to represent positive sample amplitudes, and half are used to represent negative sample amplitudes.

If we use 16 binary digits to represent the largest possible sine wave, the signal-to-noise ratio of the sine wave so produced will be about 98 dB.

Suppose that we choose 50,000 sample amplitudes a second. We then produce a sequence of 50,000 short electric pulses a second, such that the amplitude of each is equal to the sample amplitude. We then pass this train of pulses through a low-pass filter to eliminate any frequency components above 25,000 Hz. Because we are free to choose the numbers that give the sample amplitudes *in any way that we wish,* we can by this means produce *any possible* sound wave whose bandwidth is 25,000 Hz or less, at least with the accuracy indicated. This is good enough for high-quality musical sounds.

This is too much freedom of choice to be of any use. In Music V Mathews found a way to program a computer to produce a wide variety of musically useful sounds. Indeed, subsequent experience seems to indicate that Music V can be used to produce sounds as elaborate as we wish, including the sound of a singing voice and even speech sounds.

How is this done in Music V? We have all heard of the idea of a computer *simulating* some sort of mechanical or electrical device. The easiest way for noncomputer people to understand Music V is to think of it as making the computer simulate the operation of several fundamental electronic devices that are connected together in various ways. These fundamental devices are shown in the upper figure on the facing page.

One important device is the *oscillator,* represented in part A. This has an output and two inputs. The number going into input I1 specifies the amplitude of the output wave; the number going into I2 specifies the frequency of the output wave. Each oscillator is programed to produce a particular waveform *Fn,* which may be a sine wave, a square wave, or some other wave.

The *adder* shown in part B is essential. The output of the adder is the sum of the two inputs, I1 and I2. We may use the adder to add the outputs of two sinusoidal oscillators to get the sum of two partials. Or we may use the adder to add a small sinusoidal vibrato to the number that specifies the average frequency of the oscillator.

The output of the *multiplier* shown in part C is the product of the two input numbers I1 and I2. The multiplier may be used in several ways: to multiply by some chosen number the amplitudes of all outputs produced by an oscillator, so that we can use one number as a volume control; or to change the frequency of an oscillator by a con-

stant factor, thus transposing any notes played.

The final device (part D) is an output device that stores the sequence of numbers that represent the samples of the waveform produced. This storage may be accomplished in computer memory, on disk, or on tape.

The lower figure on the facing page shows how to program an *instrument* and play two notes. This instrument makes use of two oscillators. Oscillator F2 produces the "squarish" waveform shown in part C. Oscillator F1 produces a single time-varying output that rises from zero and falls to zero again, as shown in part B. This output determines how the amplitude of the output of oscillator F2 rises and falls with time. Thus, the amplitude and duration of the note produced by F2 are controlled by the input numbers P5 and P6. The frequency of the note is controlled by input P7.

The program for "creating" the instrument and for playing the two notes shown in part D is given in lines 1 through 10 of the figure. In this program, lines 1 through 5 define the instrument shown in part A. Line 6 defines the time function (part B) that oscillator F1 produces. Line 7 defines the waveform (part C) that oscillator F2 produces. Lines 8 and 9 cause the instrument to play the two notes shown in part D.

In line 8, the 0 following NOT says to start this note at a time zero. The 1 that follows says that the note is to be played by instrument number 1 (defined in lines 1 through 4). The 1000 that follows specifies the amplitude of the output. The following number, .0128, is, in fact, the input number P6 of part A. When this number times the number of successive samples is equal to 511, the output of oscillator F1 will have risen from zero and fallen to zero again, as shown in part B and line 6. The first note is two seconds long, and it is assumed in the example that there are 20,000 samples per second. The entry for P6 in line 8 is .0128. We note that

$$(.0128)(2)(20,000) = 512$$

which is close enough.

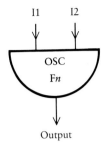

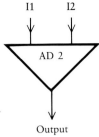

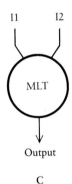

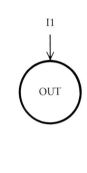

**A**

**B**

**C**

**D**

■ The four devices that Music V "simulates." Part **A** represents an oscillator *Fn* whose frequency is controlled by input number I2 and whose output amplitude is controlled by input number I1; part **B**, an adder whose output number is the sum of the two input numbers I1 and I2; part **C**, a multiplier whose output number is the sum of the input numbers I1 and I2; and part **D**, an output device to store a sequence of numbers that represent sample amplitudes.

■ A simple instrument, and the program that causes it to play the two notes shown in part **D**. The instrument consists of an oscillator, F1, that produces the envelope (part **B**) of the output of oscillator F2. The waveform of oscillator F2 is shown in part **C**. Lines 1 through 5 describe the instrument; lines 6 and 7 describe the waveform; and lines 8 through 10 play the notes.

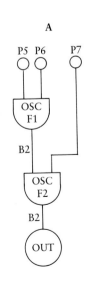

**A**

**B**

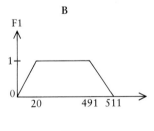

**C**

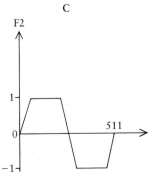

**D**

**E**

1 INS 0 1 ;
2 OSC P5 P6 B2 F1 P30 ;
3 OSC B2 P7 B2 F2 P29 ;
4 OUT B2 B1 ;
5 END ;
6 GEN 0 1 1 0 0 .99 20 .99 491 0 511 ;
7 GEN 0 1 2 0 0 .99 50 .99 205 −.99
   306 −.99 461 0 511 ;
8 NOT 0 1 2 1000 .0128 6.70 ;
9 NOT 2 1 1 1000 .0256 8.44 ;
10 TER 3 ;

Square
waveform

Sawtooth
waveform

■ Two simple waveforms that can be produced by using Music V.

The last number in line 8 is the input P7, and this specifies frequency. According to part C and line 7, one complete cycle happens when P7 times the number of samples equals 511. Hence, the number of cycles per second, or Hz, is P7 times 20,000, so that the frequency is

$$(6.70)(20,000)/(511) = 262 \text{ Hz}.$$

This is indeed the frequency of middle C.

The NOT lines of the figure are a very primitive way of playing a very primitive instrument. The next-to-last number, P6, depends on note duration. Surely the computer can compute this for us! Indeed, it can and does. If we wish, we can enter the frequency instead of P7, and the computer can compute the required P7. Or we can enter the name of the note and its octave number, or the octave number and a number of semitones. We can have much more complicated instruments that make it possible to have vibrato, to sweep the frequency smoothly with time, to do a host of things.

Let us just take it for granted that we can do anything we want with Music V. What do we *want* to do?

There are several methods of sound synthesis. The most direct is to add together a lot of sinusoidal partials whose amplitudes rise and fall somewhat differently for the duration of the note. This is called *additive synthesis*. It is very powerful but somewhat slow, because each sinusoidal partial must be computed separately.

In the early days of sound synthesis, the oscillators of Music V were programmed to produce geometrically simple waveforms, such as those shown in the figure above. This was economical, but the sounds produced were limited and inflexible in quality, and not very good.

In his book *The Technology of Computer Music*, Mathews describes an instrument whose waveform varies with amplitude (see the figure at the right). Such an instrument has features that ordinary musical instruments have: the quality of the tone changes as the intensity of the tone is in-

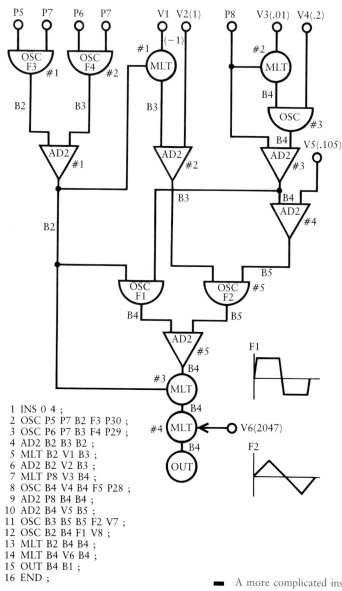

```
1 INS 0 4 ;
2 OSC P5 P7 B2 F3 P30 ;
3 OSC P6 P7 B3 F4 P29 ;
4 AD2 B2 B3 B2 ;
5 MLT B2 V1 B3 ;
6 AD2 B2 V2 B3 ;
7 MLT P8 V3 B4 ;
8 OSC B4 V4 B4 F5 P28 ;
9 AD2 P8 B4 B4 ;
10 AD2 B4 V5 B5 ;
11 OSC B3 B5 B5 F2 V7 ;
12 OSC B2 B4 F1 V8 ;
13 MLT B2 B4 B4 ;
14 MLT B4 V6 B4 ;
15 OUT B4 B1 ;
16 END ;
```

■ A more complicated instrument, whose waveform depends on the amplitude, and the program that defines it.

A

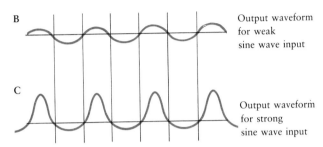

B — Output waveform for weak sine wave input

C — Output waveform for strong sine wave input

■ A waveform that changes with amplitude can be produced by using a sine wave as an input to a device with the input-output characteristic shown in part **A**. If the input is a sine wave of small amplitude, the output will be almost sinusoidal, as shown in part **B**. If the input is a sine wave of larger amplitude, the output will be a peaked waveform, as shown in part **C**; this has many harmonic partials.

■ Chowning's fm synthesis. At the top is a sine wave whose amplitude is a function of a variable $t'$. Part **B** shows the rate at which $t'$ changes with time, $t$. This rate has a sinusoidal variation with time, a sort of vibrato. In part **C**, the waveform of part **A** is plotted, not against $t'$, but against actual time $t$. The waveform is peaked and has many harmonic partials. By gradually increasing the amount by which $t'$ varies with time, we can gradually increase the strengths of the higher partials.

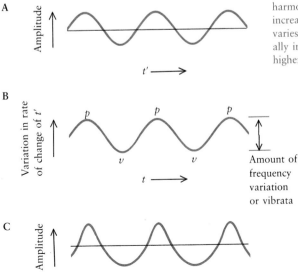

creased, and louder tones have more partials than weaker tones. The adjacent figure illustrates what can be accomplished with an instrument of this sort. Curve A is a plot of the nonlinear relation between the output amplitude and the input amplitude. Waveform B shows the output waveform when the input is a weak sine wave; the output waveform is almost sinusoidal and has chiefly one partial. As shown in part C, the output for a strong sinusoidal input is peaked, and has many harmonic sinusoidal partials: the stronger the input, the stronger the higher partials. Various computer musicians have used this means to obtain rich tones whose timbres change with amplitude.

John Chowning invented another method, which easily produces rich, brassy sounds, called *fm* (frequency modulation) *synthesis*. It is done, in effect, by putting a sinusoidal vibrato on a sinusoidal oscillator, but the vibrato has the *same frequency as the oscillator*. In part A of the adjacent figure, the amplitude of the output is plotted against a sort of pseudotime $t'$, which governs the rate of generation of samples. Curve B shows the effect of the vibrato on the rate of change of $t'$ with actual time $t$. At points $p$, $t'$ changes rapidly with $t$; at points $v$, $t'$ changes slowly with $t$. The resulting output waveform is shown in part C. The peaks rise and fall rapidly; the valleys fall and rise slowly. The resulting waveform has many high-frequency harmonics. Moreover, the intensities of these high-frequency partials can be increased by increasing the amount of frequency modulation (vibrato). If we gradually increase the amount of frequency modulation during the playing of a note, the intensities of the higher partials gradually increase. This is characteristic of trumpet tones. Of course, the amplitude of the sine wave that is frequency modulated in this way should rise and fall during the playing of the note.

Other interesting effects can be obtained by frequency modulating with a frequency that is an integer multiple of the oscillator frequency, or with a frequency that is not quite equal to the frequency of the oscillator.

Amplitude of partial →

$f_0 = nf_p$

Frequency →

216

Chowning has made another interesting use of frequency modulation. In the sound spectrum in the figure above, the "trick" is to use a frequency $f_p$ to frequency-modulate an oscillator with a frequency $f_0 = nf_p$. Here $n$ is an integer, such as 17 or 25. As shown in the figure, this creates partials whose frequencies are all integer multiples of $f_p$ and whose peak amplitude is at $f_0$. This spectrum is much like that of the formant of a voice sound, which consists of partials of a pitch frequency $f_p$ centered on some formant frequency $f_0$. Chowning has used this technique to synthesize very good singing voice sounds made up of several formants produced in this way.

In *subtractive synthesis* one starts out with some waveform that has many partials, such as those in the figure at the top of page 214, and, in effect, passes them through a (digital) filter. This apes the actual way in which the voice and various musical instruments produce sound. Subtractive synthesis takes a lot of computer time, and the results seem no better than (and perhaps inferior to) other methods of synthesis. Experience seems to indicate that it is better to work toward the desired waveform or spectrum of a sound than to try to simulate the physical behavior of mechanical sound sources.

One important feature of sound generation is the use of computer-produced random or pseudorandom noise. This is useful in making a vibrato or tremolo not quite mechanically regular. Also, band-limited or band-shaped noise (analogous, in a way, to a whisper) has been used in some compositions. There are other "tricks" of synthesis besides those described here. One is the *ring modulator,* which shifts all partials up by a constant additional frequency.

Although computer generation of sound is in principle the most powerful and flexible way to produce new sounds, computer processing of sound has found an increasing place in music. The computer may be used merely to record, combine, splice, speed up, or slow down sound, to play it backward, or to filter it to modify its spectrum. All of this had been done before in *Musique concrète* by splicing magnetic tape, playing it at speeds other than that at which it had been recorded, or playing it backward, or filtering the recorded sounds. This is very laborious with tape, and noise accumulates during re-recording. At Bell Laboratories and in other places, the computer has been used to edit and assemble various recorded sounds, as an alternative to splicing.

The computer can be used to do much more than such simple manipulations of sound. As noted in Chapter 3 and in Chapter 13, two important characteristics of periodic sounds, such as the voice and most musical sounds, are the periodicity or pitch frequency, which gives the pitch, and the spectral envelope, the amplitude of partials as a function of frequency, which greatly affects the timbre. In vowel sounds the pitch frequency determines the pitch, and the spectral envelope determines which vowel sound is heard.

We have noted that a computer can be used to analyze sounds to identify both the pitch frequency and the spectral envelope. Using data so derived, one can synthesize a sound with the original spectral envelope but with a different pitch frequency. One can also modify the spectral envelope in various ways. In *Lions Are Growing,* Andy Moorer thus derived from two spoken versions of Richard Brautigan's poem a singing voice, singing in chords and appropriate roaring sounds.

■ If a sine wave of frequency $f_p$ modulates the frequency of a sine wave of frequency $f_0 = nf_p$, in which $n$ is an integer, one obtains a wave whose spectrum consists of many partials centered on the frequency $f_0$. These partials are all harmonics of the frequency $f_p$. This resembles the spectrum of a formant of a vocal sound.

# Appendix G  Microbiographies

I didn't want to interrupt the flow of the text by identifying in detail all those persons mentioned. Some are in essence identified by association with others. Some need no identification (Bach, Beethoven). But others, historical and current, merit further discussion here.

**Babbitt, Milton (1916–    ).** Composer, theorist, and proponent of total serialism. Professor of Music at Princeton University, and sometime director of the Columbia-Princeton Electronic Music Center, where he worked closely with Otto Luening and Vladimir Ussachevsky. Among his compositions are *All Set* (a mathematical pun) and the fine *Philomel* for synthesizer and soprano.

**Batteau, Wayne (1916–1967).** An independent researcher who pointed out the importance of the pinna (outer ear) in sensing the height of sound sources.

**Békésy, Georg von (1899–1972).** Received the Nobel Prize for Physiology or Medicine in 1961 for his discovery of the functioning of the cochlea and its basilar membrane. His book *Experiments in Hearing* (1960) is in print.

**Boulez, Pierre (1925–    ).** French composer, pianist, and conductor. Boulez is the director of IRCAM (Institute for Research and Coordination of Acoustics and Music, attached to the Pompidou Center in Paris), which owes its existence to him. The extent and variety of his musical activities are immense.

**Broadbent, Dennis E. (1926–    ).** Distinguished British psychologist, author of *Perception and Communication*, Pergamon Press, 1958. For many years Broadbent was director of the Applied Psychology Unit of the Medical Research Council, at Cambridge. He is now Professor of Experimental Psychology at Oxford.

**Cage, John (1912–    ).** American composer and mycologist who has used chance in composing some of his pieces. Among his best known works is *4′ 33″* (of silence).

**Cagniard de la Tour, Charles (1777–1859).** French engineer. Inventor of the siren.

**Chavez, Carlos (1899–1978).** Mexican composer and conductor. His *Toccata for Percussion* (1942) is scored for eleven types of percussion instruments, some of them native. In part, it was composed to exhibit the sounds produced by devices used by early Mexican Indians.

**Chowning, John M. (1934–    ).** Chowning's degrees are a B. Music in composition, Wittenberg University (1959), and a Ph.D. in composition from Stanford University (1966). With some help from Max Mathews, in 1964 Chowning set up a computer music program at Stanford University. He has made Stanford the leading center for computer-related musical activities. His compositions explore new areas of musical effect: *Turenas,* sound in motion; *Stria,* nonharmonic partials; *Phōnē,* voicelike sounds and their transformations.

**Cohen, Elizabeth Ann,** without whose help this book would probably not exist, studied at the University of California at Berkeley, at Bennington College (B.A. in music and science, 1975), and at Stanford University (M.S. in electrical engineering, 1978; Ph. D. in acoustics, 1980) and worked at Bell Laboratories during the 1970s. She is expert with the computer and the library.

**Cowell, Henry (Dixon) (1897–1965).** A prolific American composer who wrote in many styles. In his early piano and orchestral music he used tone clusters, which he regarded as chords, contrary to views expressed in this book. Cowell's later works were increasingly tonal.

**d'Alembert, Jean Le Rond (1717–1783).** French mathematician, scientist, and philosopher, permanent secretary to the French Academy. Author of *Eléments de musique* (1752) and expositor of Rameau's ideas.

**Fletcher, Harvey (1884–1981).** An outstanding physicist who, at Bell Laboratories, led a wide range of important work on speech, hearing, and sound reproduction. His books, *Speech and Hearing* (1929) and *Speech and Hearing in Communication* (1953) summarize in a very accessible manner a wealth of results obtained at Bell Laboratories by the use of vacuum-tube electronics and good microphones, headphones, amplifiers, and loudspeakers.

**Fux, Johann Joseph (1660–1741).** Austrian composer and musicologist. His *Gradus ad Parnassum* (1725) was a textbook to Mozart, Haydn, and Beethoven. Fux's views of counterpoint were deduced from the works of an earlier generation of composers, including Palestrina (1525–1642).

**Gilbert, Edgar N. (1923– ).** A versatile mathematician at Bell Laboratories. His unique work on a search for *The Lost Chord* has never been published. Gilbert used combinatorial methods to reduce all possible chords to a listenable number. After playing all of these he felt that, although the diminished seventh was real gone, it did not fit the description given by Sullivan, and he concluded that the lost chord must be silence.

**Grainger, Percy (1882–1961).** An Australian pianist and composer. One of the first to introduce tone clusters as a musical device (as opposed to the imitation of cannon fire).

**Green, David M. (1932– ).** A very distinguished and reliable expert in all that pertains to hearing. He has served on the faculties of Harvard and the University of California at San Diego and is now at MIT. His book, *An Introduction to Hearing* (1976), is a most reliable reference.

**Grey, John M. (1947– ).** At Stanford University, Grey has made invaluable studies of musical timbre, using the computer to analyze and synthesize sounds and using multidimensional scaling to exhibit relationships among them.

**Harris, Cyril (1917– ).** Leading American designer of successful concert halls; Professor of Electrical Engineering and Architecture at Columbia University. His books, *Acoustical Designing in Architecture* (with Vern O. Knudson), 1958, reprinted 1978, and *Handbook of Noise Control*, 1980, are invaluable.

**Helmholtz, Hermann Ludwig Ferdinand von (1821–1894).** One of the greatest scientists of the nineteenth century. A physician, anatomist, physiologist, and physicist, with an insightful knowledge of music. His *On the Sensation of Tone as a Physiological Basis for the Theory of Music* presents the first coherent, detailed overview of his and others' work on the nature and perception of musical sound.

**Hiller, Lejarin A., Jr. (1924– ).** Composer and teacher; now at the Department of Music, State University of New York at Buffalo. In 1957, while at the University of Illinois at Urbana, he published, with L. M. Isaacson, a chemist with a knowledge of computers, *The Illiac Suite for String Quartet*, a computer-composed piece based on chance and the laws of first-species counterpoint. The Illiac was a computer at the University of Illinois.

**Ligeti, György (1923– ).** His compositions have been influenced by his experience at the West German radio studio for electronic music in Cologne in 1958, where Stockhausen had begun his work in 1953. There Ligeti produced some electronic pieces, of which only *Articulation* survives. Ligeti was not satisfied with the electronic sounds then available, but his work in Cologne gave him a deep interest in the details of orchestral sound, which is apparent in his later music. Ligeti served as visiting lecturer at Stanford in 1972.

**Luening, Otto (1900– ).** Composer, conductor, flautist, sometime codirector of the Columbia-Princeton Electronic Music Center.

**Mathews, Max V. (1926– ).** To whom this book is dedicated. Mathews's computer programs, culminating in Music V, made the computer accessible as a musical instrument. Many important musicians with a bent for the computer have worked with him, including Jean-Claude Risset, John Chowning, Barry Vercoe, Gerald Strang, Andrew Moorer (now at Lucasfilm), and Richard Moore (now at the University of California at San Diego). At the inception of IRCAM, and for a number of years thereafter, Mathews served as its scientific advisor. Mathews's musical activities are wide ranging. He has produced excellent electronic stringed instruments and many other ingenious devices. He is director of the Acoustical and Behavioral Laboratory at Bell Laboratories.

**Mayer, Alfred M. (1836–1897).** In 1876 Mayer published an important paper, "Researches in Acoustics," in the *Philosophical Magazine*. This paper describes masking, and asserts that a high-pitched sound cannot mask a sound of lower pitch. Mayer applies his findings to his experience with music, and notes: "In a large orchestra I have repeatedly witnessed the entire obliteration of all sounds from violins by the deeper and more intense sounds of wind instruments."

**Mersenne, Marin (1588–1648).** French mathematician, natural philosopher, and theologian. His *Harmonie universelle* (1636–1637) related pitch to periodicity at the same time that Galileo did. Mersenne was the first to measure the velocity of sound, somewhat inaccurately.

**Moorer, James A. (1945– ).** Trained in music, electrical engineering, applied mathematics, and computer science, Moorer is extremely bright and productive. He has made important contributions to computer analysis of sounds, to artificial reverberation, and to the computer transformation or processing of natural sounds, such as the human voice. He has composed very attractive music. Now at Lucasfilm in San Rafael, California.

**Penderecki, Krzysztof (1933– ).** Penderecki is noted particularly for moving compositions on a large scale, such as his *St. Luke Passion* and *Paradise Lost*, but he has produced works of great variety. Here we are concerned with his orchestral sound, which, among more conventional elements, has unusual features that seem to derive from early electronic sounds.

**Ramo, Simon (1913–   ).** Public figure in engineering and science, vice-chairman of the board of TRW (once Thompson Ramo Wooldridge). A classmate of mine at Caltech and an expert violinist, active in music in Los Angeles. He had a brief encounter with computer music in 1963.

**Risset, Jean-Claude (1938–   ).** Risset's background was perfect for computer music. In France he studied piano with Trimaille and Goullon, and composition with Demarquez and Jolivet. He worked for three years with Max Mathews at Bell Laboratories to develop the resources of computer sound synthesis. In 1967 he received his Doctorat ès-Sciences Physiques from the Ecole Normale Supérieure. He wrote a catalog of computer-synthesized sounds in 1969. He set up computer sound systems at Orsay (1971), at Marseille-Luminy (1974), and at IRCAM, where Boulez asked him to head the Computer Department (1975–1979). He is now professor at the University of Aix-Marseille, and works on computer music at Luminy and CNRS. His *Songes* received the first prize for digital music at the eighth International Electroacoustic Music Competition at Bourges, 1980, and in 1981 he was awarded the "Grand Prix de la promotion de la symphonique" by SACEM, the French society of authors.

**Sabine, Wallace Clement (1868–1919).** Hollis Professor of Mathematics and Natural Philosophy at Harvard, who founded the science of architectural acoustics. His *Collected Papers on Acoustics* was first published after his death, in 1922.

**Schouten, Jan (1910–1980).** An ingenious and engaging Dutch physicist who, for many years from its founding, was head of the Institute for Perception Research in Eindhoven, a laboratory that derived support from Philips, where Schouten had previously worked. Schouten discovered the important phenomenon of residue pitch.

**Schroeder, Manfred (1926–   ).** A powerful contributor to a wide range of acoustical science. The Third Physical Institute of the University of Goettingen, of which he is director, is *the* place in architectural acoustics. At Bell Laboratories, Schroeder heads a department, and directs work on speech synthesis and coding and on various psychoacoustic problems.

**Shepard, Roger (1929–   ).** A very distinguished psychologist whom I knew well at Bell Laboratories. He later went to Harvard and is now at Stanford. He devised multidimensional scaling, a computer-oriented nonlinear improvement over factor analysis. Shepard and his students have shown a deep interest in music. Shepard devised the ever-ascending *Shepard's Tones*.

**Sperry, Roger (1913–   ).** Nobel Laureate and distinguished physiologist and psychologist. His work and that of his followers with animals and with patients in whom the two hemispheres of the brain have been surgically isolated from one another (in human beings to cure intractable epilepsy) gave us our first sure knowledge of the capabilities residing in the two hemispheres.

**Stevens, S. S. (1903–1973).** Smitty Stevens was a charming and distinguished psychologist at Harvard for many years. Because of his persistence the power-law relation between perceived loudness and sound intensity was finally adopted as an international standard, and power laws for other modalities were adopted as well.

**Stockhausen, Karlheinz (1928–   ).** Stockhausen is well known for electronic works produced at the West German radio studio for electronic music in Cologne, which he helped to found in 1953, and of which he became director in 1963. He has written more for conventional instruments, with a great emphasis on sound texture. Some of his scores are very unusual in appearance; some give only broad directions for what is to be played, others give meticulous directions in unusual notation.

**Strang, Gerald (1908–   ).** Assistant to Arnold Schoenberg, student of Ernst Toch, editor of *New Music*, head of the music departments at California state universities at Long Beach and Northridge, Strang has composed important works for both conventional instruments and the computer.

**Sundberg, Johann (1936–   ).** Professor of Musical Acoustics, Royal Institute of Technology, Stockholm, Sweden. Sundberg is perhaps best known to musicians for his discovery and interpretation of the "singer's formant" (see Chapter 9). His research and publications have covered a wide range of musical phenomena, including: the "grammar" or organization of Swedish folk tunes and the computer generation of tunes with the same character; investigations of deviations from the conventional 2-to-1 frequency ratio in the judgment of musical octaves; investigations into the musical intervals actually played and sung, including the remarkably consistent intervals of good barbershop quartets, architectural acoustics, and other matters pertaining to music and its performance and appreciation.

**Tenney, James (1934–   ).** A talented composer and music theorist who spent some years at Bell Laboratories in the early days of computer music. He was later at Harvard and at the California Institute of the Arts, and is now at York University in Canada.

**Terhardt, Ernst (1934–   ).** An expert in musical acoustics. In 1977 in a poster demonstration at IRCAM, he made Rameau's fundamental bass clearly present by using it to play a tune. Terhardt is noted for his theory of pitch perception.

**Ussachevsky, Vladimir (1911–   ).** With Milton Babbitt and Otto Luening, he served as director of the Columbia-Princeton Electronic Music Center. Of these three composers, Ussachevsky was the only one to embrace the computer. He later went to the University of Utah, a hotbed of computer science.

**Varèse, Edgard (1883–1965).** Varèse destroyed much of his early music. The few works of his that survive speak for themselves. Varèse seemed unfettered by theories or preconceptions. Personally, he was charming, opinionated, and unforgettable. His wife, Louise, wrote *Varèse: A Looking Glass Diary* (1972).

**Vercoe, Barry Lloyd (1939–   ).** The free-wheeling knight of the keyboard who heads computer-music work in the Department of the Humanities at the Massachusetts Institute of Technology. Vercoe has a B. Music and a B.A. in mathematics from the University of Auckland, and a doctorate in composition from the University of Michigan.

**Wessel, David L. (1942–   ).** An experimental psychologist at Michigan State University, now at IRCAM, who has done extensive work on musical timbre, using multidimensional scaling.

**Xenakis, Yannis (1922–   ).** Born in Romania of Greek extraction, Xenakis studied at the Athens Institute of Technology and worked with the architect Le Corbusier for twelve years, collaborating with him on the design of the Philips Pavilion at the 1958 Brussels World's Fair (Varèse composed electronic music that was played in the pavilion). In 1952 Xenakis turned to music. He became critical of serial (12-tone) music, asserting that its linear organization is not heard in complex polyphony, where the ear is chiefly sensitive to densities of sound that are not controlled by the twelve-tone organization. Instead, Xenakis controls the overall organization and the broad features of his compositions statistically (on the average) rather than controlling all small details. The mathematics in his *Formalized Music* seems disjointed and sometimes irrelevant to this engineer, and would throw any nonmathematical musician off, but it is apparently of use to Xenakis, for he composes very attractive music.

Apel, Willi. *Harvard Dictionary of Music.* Harvard University Press, 1958. A standard reference work.

Bekesy, Georg von. *Experiments in Hearing.* McGraw-Hill, 1960.

Bliven, Bruce. "Annals of Architecture," *New Yorker,* Nov. 8, 1976.

Broadbent, Donald E. *Perception and Communication.* Pergamon Press, 1958.

*Computer Music Journal.* Published by MIT Press. Particularly appropriate for readers of this book, but expensive.

Damaske, P. "Head-Related Two-Channel Stereophony with Loudspeaker Reproduction," *J. Acoustical Soc. Amer.* 50(1971):1109–1115.

*dB: The Sound Engineering Magazine.* Useful for those who want to keep up with sound reproduction blow by blow. Available from Sagamore Publishing, 1120 Old Country Rd., Plainview, Long Island, NY 11803.

Deutsch, Diana, ed. *The Psychology of Music.* Academic Press, 1982. This book is made up of excellent sections on various musical topics and gives copious references.

*Encyclopaedia Britannica.* The general articles on music in all editions are accurate and useful.

Fletcher, Harvey. *Speech and Hearing.* Van Nostrand, 1929.

————. *Speech and Hearing in Communication.* Van Nostrand, 1953.

Gilbert, E. N. "An Iterative Calculation of Reverberation," *J. Acoustical Soc. Amer.* 69(1981):178–184.

Green, David M. *An Introduction to Hearing.* Lawrence Erlbaum Assoc., 1976. The clearest and best-balanced guide to sound and hearing that I know. It includes all the theories and quibbles that I have deliberately omitted from this book and gives excellent references to the technical literature.

Grey, J. M. "Multidimensional Perceptual Scaling of Musical Timbre," *J. Acoustical Soc. Amer.* 61(1977):1270–1277.

Hall, Donald E. *Musical Acoustics.* Wadsworth, 1980. Perhaps the best available textbook.

Harris, Cyril. *Handbook of Noise Control,* 2d ed. McGraw-Hill, 1979.

Helmholtz, Hermann von. *On the Sensation of Tone.* Dover reprint of 2d English ed., 1954. This is a book to browse through. It is the basic book in acoustics.

Hutchins, Carleen Maley. *The Physics of Music.* W. H. Freeman and Company, 1978. Discusses in detail the mechanics of musical instruments and the human voice, which I do not.

*Journal of the Acoustical Society of America.* A major source of professional information and research results in acoustics.

Knudsen, Vern D., and Cyril M. Harris. *Acoustical Designing in Architecture.* American Institute of Physics reprint, 1978. Both valuable and a bargain.

Mathews, Max V. *The Technology of Computer Music.* MIT Press, 1969. Invaluable to anyone who wishes to pursue computer generation of sound or to understand it in depth.

Mathews, Max V., and John R. Pierce. "Harmony and Nonharmonic Partials," *J. Acoustical Soc. Amer.* 68(1980):1252–1257.

Mayer, Alfred M. "Researches in Acoustics," *Philosophical Magazine,* 1876. Reprinted in *Psychological Acoustics,* Earl D. Schubert, Dowden, Hutchinson, and Ross, 1979.

Mellert, V. "Construction of a Dummy Head after New Measurements of the Threshold of Hearing," *J. Acoustical Soc. Amer.* 51 (1972):1359–1361.

Olson, Harry F. *Modern Sound Reproduction.* Robert E. Krieger reprint, 1978. The classic book in the field of sound reproduction. Technologically out of date, but contains useful background material.

222

Pierce, J. R. "Attaining Consonance in Arbitrary Scales," *J. Acoustical Soc. Amer.* **40**(1966): 249.

Plomp, R. *Aspects of Tone Sensation*. Academic Press, 1976. An important book by an important researcher.

Plomp, R., and W. J. M. Levelt. "Tonal Consonance and Critical Bandwidth," *J. Acoustical Soc. Amer.* **38**(1965):548–560.

Rameau, Jean-Phillipe. *Treatise on Harmony*. Dover reprint, 1971. An important book, but very hard going. I have discussed most of what is relevant in it in this book.

*Recording Engineer Producer*. A major professional journal for sound reproduction. Available from Callay Communications, 1850 North Whitley Ave., Hollywood, CA 90028.

Risset, Jean-Claude, and Max V. Mathews. "Analysis of Musical Instrument Tones," *Physics Today* **22**(1969):23–30.

Roads, C., J. Snell, and J. Strawn, eds. *Computer Music*. MIT Press, 1982. Contains a wealth of relevant information on acoustics.

Rossing, Thomas D. "Acoustics of Percussion Instruments, Parts I and II," *The Physics Teacher* **14**(1976):546–556 and **15**(1977): 278–288.

———. "Physics and Psychophysics of High-Fidelity Sound, Parts I–IV," *The Physics Teacher* **17**(1979):563; **18**(1980):278 and 426; **19**(1981):293–304.

Rossing, Thomas D., and H. John Sathoff. "Modes of Vibration and Sound Radiation of Handbells," *J. Acoustical Soc. Amer.* **68** (1980):1600–1607.

Sabine, Wallace Clement. *Collected Papers on Acoustics*. Harvard University Press, 1922. Dover reprint, 1964.

Schouten, J. F. "The Residue, a New Component in Subjective Sound Analysis," *K. Ned. Akad. Wet. Proc.* **43**(1940):356–365.

Schroeder, M. R. "Computer Models for Concert Hall Acoustics," *Amer. J. Physics* **41**/4(1973):461–471.

———. "Toward Better Acoustics for Concert Halls," *Physics Today* **33**(1980):24–30.

Schroeder, M. R., B. S. Atal, G. M. Sessler, and J. E. West, "Acoustical Measurements in Philharmonic Hall (New York)," *J. Acoustical Soc. Amer.* **40**(1966):434–440.

Schroeder, M. R., and D. Hackman. "Iterative Calculation of Reverberation Time," *Acoustics* **45**(1981):269–273.

Schubert, Earl D., ed. *Psychological Acoustics*. Dowden, Hutchinson, and Ross, 1979. The original papers reprinted here, which date from 1876 to 1970, give a wonderful sense of how the discoveries in acoustics were really made, and the wise introductory comments to the six sections give references up to 1977.

Shankland, Robert S. "Acoustical Designing for Performers," *J. Acoustical Soc. Amer.* **65** (1979):140–141.

Shepard, Roger. "The Analysis of Proximities: Multidimensional Scaling with an Unknown Distance Factor," *Psychometrics* **27**(1962): 125–140, 219–246.

Slaymaker, Frank H. "Chords from Tones Having Stretched Partials," *J. Acoustical Soc. Amer.* **47**(1970):1569–1571.

Sundberg, Johann. "The Acoustics of the Singing Voice," *Scientific American* **236** (1977):82–91.

Terhardt, Ernst. "Pitch, Consonance, and Harmony," *J. Acoustical Soc. Amer.* **55** (1974):1061–1069.

Terhardt, Ernst, Gerhard Stoll, and Manfred Seewann. "Pitch of Complex Signals According to Virtual Pitch Theory," *J. Acoustical Soc. Amer.* **71**(1982):671–678.

Van Bergjeick, William A., John R. Pierce, and Edward E. David, Jr. *Waves and the Ear*. Doubleday, 1960. Tells much more about the neuroanatomy and neurophysiology of hearing than I have dealt with here, but is unfortunately now somewhat out of date.

Wegel, R. L., and C. E. Lane. "The Auditory Masking of One Pure Tone by Another and Its Probable Relation to the Dynamics of the Inner Ear," *Physical Review* **23**(1924):266–285.

Woram, John. *The Recording Studio Handbook*. Sagamore, 1977. Useful information on sound reproduction.

Xenakis, Yannis. *Formalized Music*. Indiana University Press, 1971.

# Appendix I  Description of the Recorded Musical Sound Examples

Computer sound synthesis affords the user precise control of the physical structure of the synthesized sounds. This permits the manufacture of sounds with great accuracy and reproducibility, thus yielding musical possibilities that are not available from other sound sources.

The process demands that the user provide a complete physical description of the desired sound, whereas the musician is really concerned with the auditory effect of the sound. Hence, psychoacoustics is essential in taking advantage of the wide possibilities of computer synthesis (and computer synthesis of sound is an essential tool of psychoacoustics).

The examples on the two records with this book attempt to illustrate this point. Each example is explained in some detail here, so that the listener can both experience the auditory effect of the sounds and know how these sounds were constructed. The processes involved are related to topics discussed in the body of this book and are intended to illustrate some aspects of the relationship between the physical structure and the auditory effect of sound.

Some of the examples were synthesized by Max Mathews and myself at IRCAM in 1979. Many were created by Elizabeth Cohen at Stanford University, with the cooperation and advice of John Chowning and the assistance of Andrew Schloss and David Jaffe. Diana Deutsch allowed me to use recordings of some of her musical paradoxes, and Max Mathews allowed me to use his brasslike violin sounds. The rest of the examples were synthesized by Jean-Claude Risset, using Max Mathew's Music V program, mostly in Marseille (at the Laboratoire d'Informatique et Acoustique Musicale, Faculté des Sciences de Luminy, et Laboratoire de Mécanique et d'Acoustique, Centre National de la Recherche Scientifique) and a few at Bell Laboratories and IRCAM.

## Examples for Chapter 3, Sine Waves

Sine waves, known in music as pure tones, are important both to music and to the study of hearing, for mathematical, physiological, and perceptual reasons. However, we must remain cautious when we try to extrapolate from the perception of sine waves to the perception of music.

*Example 1.1. Sine waves of two different frequencies.*

The perceived quality of a sine wave (pure tone) changes with its frequency. The first tone, whose frequency is 80 Hz, sounds like a dull hum. The second tone, of 1,000 Hz, sounds like a dull whistle. Both tones have the same physical intensity, but the 1,000-Hz tone sounds louder.

*Example 1.2. Construction of a square wave by adding sinusoidal components.*

A perfect square wave is made up of an infinite number of harmonically related sine waves of the proper amplitudes and phases. The frequencies of these sinusoidal components or partials must be odd multiples (harmonics) of the fundamental frequency. The figure at the top of page 43 shows how a square wave can be approximated better and better by adding more and more harmonics. If the phases of the partials are not correct, the sum of the frequency components will not give a square wave, but the wave sounds a good deal like a square wave. The sounds are:

a. 1st, 3rd, and 5th harmonics;
b. 1st, 3rd, 5th, 7th, 9th, and 11th harmonics;
c. 1st, 3rd, 5th, 7th, 9th, 11th, 13th, 15th, 17th, 19th, 21st, 23rd, 25th, 27th, and 29th harmonics;
d. same as part *c* but random phases.

## Examples for Chapter 4: Scales and Beats

*Example 1.3. Beats.*

When two sine waves of nearly the same frequency are sounded together, beats are produced (see the figure on page 63). Beating is heard as a wavering of loudness, a sort of trem-

The dominant seventh cadence.

olo, whose frequency is the difference of the frequencies of the two sine waves. When the beating is rapid, the sound is harsh, as in parts *b* and *c* here. When the frequencies are separated enough, we do not hear beats; we hear the two tones separately (see Chapter 5). In the examples, one tone has a frequency of 400 Hz, and the second differs by:

**a.** 2 Hz;
**b.** 15 Hz;
**c.** 20 Hz.

*Example 1.4. Just and equal-tempered tuning contrasted.*

A scale in just tuning, part *a*, is here followed by a scale in equal-tempered tuning, part *b*. Most people do not easily recognize the difference between a just and an equal-tempered scale. In part *c*, both scales are played simultaneously, and we can detect the differences in the frequencies of their notes by means of the beats that are produced. There are no beats between the beginning notes and notes an octave above, because the frequencies are equal. The beats for the fourth and fifth are slow, because fourths and fifths are about "right" in equal temperament. However, the beats between the thirds, sixths, and sevenths are faster, because the frequencies of these equal-tempered intervals are considerably "in error."

*Example 1.5. Just and equal-tempered four-note major chords (C, E, G, C).*

The chord is played (*a*) in just tuning, (*b*) in equal-tempered tuning, and (*c*) in just tuning again. The chord is harsher in equal-tempered tuning, because of beats among the harmonics of the various notes. This effect is less noticeable in real music played with traditional instruments.

*Example 1.6. A Bach fragment played with three different tunings.*

The beginning of a Bach fugue, played with (*a*) well-tempered tuning, (*b*) just intonation (Zarlino tuning), and (*c*) Pythagorean tuning.

The differences are most noticeable for the third and the sixth above the tonic.

## Examples for Chapter 5: Consonance

As a rough rule, when two sinusoidal components of a sound are less than a minor third apart, they beat or sound rough and dissonant. It is the interval size that is critical, not the frequency difference; the frequency difference (in Hz) increases for any interval as you go up the scale. For example, the octave between $A_{440}$ and $A_{880}$ spans 440 Hz, whereas the octave between $A_{880}$ and $A_{1760}$ spans 880 Hz. Musical sounds usually have many harmonic sinusoidal components. More of these fall close together in frequency for dissonant chords than for consonant chords. This is Plomp's (1964) psychoacoustic explanation for dissonance and consonance of pure tones.

*Example 1.7. Consonance and dissonance.*

If we put partials too close together, the sound is harsh and dissonant. What is too close? Judge for yourself. Here we have tones with partials covering four octaves, with partials:

**a.** an octave apart;
**b.** a half octave apart;
**c.** an octave apart;
**d.** a quarter octave apart;
**e.** an octave apart;
**f.** an eighth octave apart;
**g.** an octave apart.

*Example 1.8. Doctoring a cadence.*

An unstretched $V_7$, I cadence (see the above figure). As you can hear in part *a*, the $V_7$ chord is clearly rough (dissonant). Partials were removed from the $V_7$ chord, leaving no remaining partials closer in frequency than three semitones (a minor third). Some partials were also removed from the I chord, so that in it also no partials were closer than a minor third. The partials removed are shown in the table on the facing page. The $V_7$ chord became smooth (tonally consonant), as you can hear in part *b*, but was still

recognized as $V_7$ by the musically proficient. This indicates that whatever role tonal dissonance may have played in the development of harmony, it is not necessary in recognizing common harmonic relationships.

*Example 1.9. A musical fragment.*

This fragment takes advantage of dispersed harmonics, which enter separately.

## Examples for Chapter 6: Harmony

This section presents tones and scales that are stretched uniformly, in the sense that the octave is stretched to be some frequency ratio *A* other than 2, such as 2.05 or 2.2. Semitones are still a twelfth of a stretched "octave," and the separations of all partials of tones are stretched in accord (see Chapter 6 for details). If the partials of two unstretched tones coincide, the partials of the stretched tones will coincide also. If the partials of two unstretched tones are well separated, the partials of the two stretched tones will be well separated. If unstretched chords are consonant, in the sense that they do not display unpleasant beats or harshness, then this theory supposes that the stretched versions of the chords will be consonant also.

*Example 2.1. Stretched scales.*

Here a scale (C, D, E, F, G, A, B, C), with partials 1, 2, 3, 4, 5, 6, 8, 10, 12, 14, down 9 dB/octave, is:

**a.** played twice unstretched (*A* = 2), then
**b.** played twice stretched (*A* = 2.4).

*Example 2.2. Stretched tunes.*

Here a hymn tune, harmonized as shown in the figure below, with the same partials as the scale in Example 2.1, is played twice:

**a.** unstretched (*A* = 2);
**b.** stretched (*A* = 2.4).

This is followed by a stretched version of "Are you sleeping, Brother John?"

**Partials removed from dominant seventh cadence.**

| Chord | Note | Number of partials removed | | | | | | |
|-------|------|---|---|---|---|---|---|---|
| | | 1 | 2 | 3 | 4 | 5 | 6 | 8 |
| $V_7$ | G | | | | X | | | X |
| | B' | | | X | X | | | X |
| | D'' | | | | X | | | X |
| | F'' | | | X | X | | | |
| I | C | | | | | | | |
| | G' | | | X | | | | X |
| | C'' | | | | X | | | X |
| | E'' | | | | | X | | |

Old Hundredth, as harmonized by Gerald Bennett, May 2, 1979.

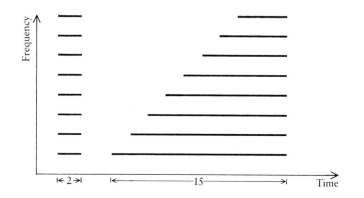

A cadence (left) and an evasive cadence (right).

■ Synthesis of a tone by the addition of partials.

*Example 2.3. Stretched cadences.*

The difference between a particular evasive cadence and a true cadence is the position of one note in an inner voice, as shown in the figure above. The difference is very clearly apparent in the unstretched versions but disappears (as well as I can hear) with a stretching of $A = 2.4$. Here we hear:

**a.** unstretched ($A = 2$) cadence;
**b.** unstretched ($A = 2$) evasive cadence;
**c.** stretched ($A = 2.4$) cadence;
**d.** stretched ($A = 2.4$) evasive cadence.

*Example 2.4. Cadences with stretched partials.*

This example illustrates more dramatically the effect of stretching partials. The familiar closing quality of a cadence is lost if we stretch the tones and intervals, even though consonant chords remain consonant. Here a conventional cadence is played with $A$ equal to:

**a.** 2.0 (unstretched, normal);
**b.** 2.05;
**c.** 2.0;
**d.** 2.2;
**e.** 2.0.

*Example 2.5. Stretching and lack of fusion.*

Unstretched tones, which have harmonic partials, have timbre: they fuse into a single sound of a certain quality. If we stretch a tone too much, so that the partials depart a great deal from being harmonics of the first partial, the tone "falls to pieces" and is heard as a mixture of high and low frequencies. Here we hear tones for which $A$ equals:

**a.** 2;      **i.** 2;
**b.** 2.05;   **j.** 2.4;
**c.** 2;      **k.** 2.
**d.** 2.1;
**e.** 2;
**f.** 2.15;
**g.** 2;
**h.** 2.2;

*Example 2.6. Fusion of partials.*

As discussed in Chapter 3, many sounds can be obtained as "combinations of sine waves or partials, even though the amplitudes and frequencies of the partials change with time." Thus harmonic synthesis manufactures the sound as a sum of harmonics; for example, waves of frequencies $f$, $2f$, $3f$, and so forth. In this example, we hear first a steady periodic tone of frequency $f_0 = 220$ Hz and of duration two seconds, followed by the gradual reconstruction of that tone by harmonic synthesis: harmonic 1 enters first, followed by harmonic 2 one second later, then by harmonic 3 two seconds later, and so on, as shown in the figure above. All harmonics have equal amplitude. Each harmonic can be clearly heard when it comes in, but later tends to fuse with the other harmonics although its amplitude stays constant; the timbre clearly changes during the synthesis.

*Example 2.7. Fusion into a chord.*

When do sinusoidal components fuse and give a unitary impression of a particular sound quality? When do they sound like different frequencies played together? Especially, when do individual sine waves fuse to give the effect of a chord?

McNabb and Chowning have discovered that the addition of a little vibrato, the same for all partials, helps to fuse the partials into a single tone. We recognize a major triad (CEG) as a familiar chord when it is played with complex tones, each of which has a number of harmonic partials. But the same chord played with sine waves (pure tones) is just the fourth, fifth, and sixth harmonic partials of a frequency two octaves below the C of the chord.

**a.** A tone whose fundamental is an octave below middle C. The note begins with all six partials and vibrato. Then the vibrato is removed; subsequently, the partials are removed, beginning with the lowest and ending with partials 4, 5, and 6.
**b.** This is simply the reverse of part *a*; that is, this begins with partials 4, 5, and 6 and no vibrato, then partials are added successively, followed by the addition of vibrato.

**Examples for Chapters 7 through 12: Power and Reproduction**

*Example 2.8. Binaural beats.*

This example is meaningful only with headphones; with loudspeakers, we get real beats between sounds from the two speakers. Here

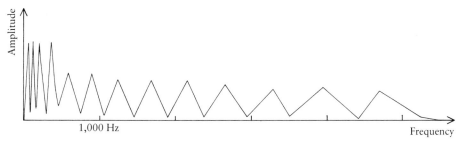

■ Envelope for synthesis of a
stringlike sound.

440 Hz is presented to one ear and 442 Hz to
the other. Most people hear a phenomenon of
sideness or beating that fluctuates at a frequency
of 2 Hz.

*Example 2.9. Power distribution.*

When a tone is made up of several well-separated harmonic partials, the loudness of the
partials add. When we merely increase the intensity of one partial, the loudness increases. Thus,
a given power distributed equally among six
harmonic partials is *much louder* than the same
power at one harmonic partial. Here we hear:

**a.**  a sine wave (pure tone); and
**b.**  same power as in part *a* divided equally
among six harmonic partials, the frequency of
the first partial being the same as in part *a*.

*Example 2.10. Effect of delay on localization.*

This example works with headphones or speakers. The sound on each track is the same, except
that the relative delays between the two tracks,
measured in milliseconds, are:

**a.**  −2;
**b.**  −1;
**c.**  −0.5
**d.**  0;
**e.**  +0.5
**f.**  +1;
**g.**  +2.

*Example 2.11. A musical fragment.*

This fragment uses the instrumentlike sounds of
which examples are given on Side 3. It is somewhat minimal, because it tends to use only G#
pitches, apart from a cymballike sound, whose
pitch is unclear.

**Examples for Chapter 13 and Appendix F: Syn-
thesis of Sounds.**

Musical sounds have pitch, loudness, duration,
and timbre. If two sounds are the same in pitch,
loudness, and duration, then whatever difference there is between them must be a difference

in timbre. What determines it? The envelope,
or way in which sound intensity rises and falls, is
important. So is the intensity of the various harmonics. So are frequency ranges in which the
harmonics are more intense (called formants). In
real musical sounds, different harmonics rise
and fall with different envelopes; their frequencies may fluctuate with time; and sometimes a
little noise is present in the sound, usually at its
beginning. Here we will explore a few of these
effects.

*Example 3.1. Stringlike tones.*

These tones have been synthesized according to
the model implemented by Max Mathews in his
electronic violin. A complex tone corresponding
to the vibration of the bowed string is modified
according to a jagged frequency response (see
the above figure). Actually, the frequency response has been made up to schematize actual
violin responses. Even with this schematized
model, vibrato tones are evocative of bowed
strings, probably because the vibrato carries
with it a complex amplitude modulation. When
the frequency goes up, a given harmonic will either increase or decrease in amplitude, depend-
ing on whether it falls on an ascending or a descending part of the frequency-response curve.

*Example 3.2. Harmonic synthesis of a brassy
tone.*

This is a complex harmonic additive synthesis.
The harmonics are controlled by different envelopes, which imitates a short and brassy tone.
The duration of the tone is .15 second; the attack and decay times range between 10 and 45
milliseconds. As indicated in the figure below,
one will hear the tone with fifteen harmonics,
followed by the first harmonic only, then first
plus second, and so on, up to the initial tone
with fifteen harmonics. There are simpler ways
to synthesize such tones, as in the next example.

*Example 3.3. Synthesis of brasslike tones.*

The brasslike tones in Example 3.2 were generated by harmonic additive synthesis; the tones
here were generated by Chowning's frequency-
modulation technique (see Appendix F). With
this technique, one can control globally the
bandwidth of the spectrum produced without
having to control the components individually.
This permits one to implement in an elegant
fashion a characteristic property of brassy tones,
namely, that the spectrum bandwidth increases
when the amplitude increases; one can simply
apply the envelope that controls the amplitude
to the parameter that controls the bandwidth of

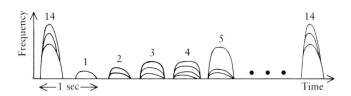

■ Harmonics of a synthesized
brassy tone.

■ Gonglike tone composed like a chord.

228

the FM spectrum (the so-called modulation index). Of the six tones presented, the second through the fifth, and especially the third and fourth, have been given some inharmonicity at the onset, to increase the "bite" of the attack. This is accomplished by making the frequency of the frequency modulation slightly different from that of the sinusoidal tone whose frequency is modulated.

*Example 3.4. Flutelike tones.*

This example gives an evocation of flute sounds rather than a close imitation. The wave contains mostly the first harmonic (or fundamental); the envelope corresponds to a gradual attack and decay, almost symmetric in time. There is some tremolo; that is, some periodic amplitude modulation. To attain a supple musical phrase and to avoid a machinelike quality, the parameters for attack times, tremolo rate, and so forth, must be varied from note to note throughout the phrase.

In the next four examples, the sound is made up of sine-wave components that are not harmonic; that is, their frequencies are not $f$, $2f$, $3f$, and so forth. The timbre depends on the frequencies as well as on the amplitudes of the components.

*Example 3.5. A gonglike tone composed like a chord.*

Bell and gong tones are inharmonic. Although the frequencies of the components are fixed in any given bell, inharmonic synthesis permits us to manufacture bell-like tones whose components have arbitrary frequencies; for example, the frequencies of the notes of a given chord. Hence, bell-like tones can be composed like chords. This example presents a chord (arpeggiated, then nonarpeggiated) played with a plucked-tone quality, followed by an inharmonic tone whose sine-wave components have the same frequencies as the notes of the chord. Then one hears the inharmonic components separately (see the above figure) as the

notes of the chord. The envelopes are restricted to a short attack (less than 10 milliseconds) and an exponential decay, which sounds like a bell or a gong if the decay is long enough. However, the sound is more natural if the lower components decay more slowly than the high ones (decay times range from 20 seconds to 2 seconds). Whether the sound is more like a bell or more like a gong depends mostly on the frequency spacing between the first components, but also on the way in which decay times vary with frequency and on the beat pattern (see Example 1.3). The gonglike tone is like an echo of the initial chord, a prolongation of the harmony into timbre.

*Example 3.6. Three steps toward a bell-like tone.*

This example shows how one can gradually improve an inharmonic synthesis to make it sound more like a bell tone. First, we hear components of frequencies $f_1$, $f_2$, $f_3$, and so forth (unequally spaced), controlled by the same envelope, giving a short attack and an exponential decay. The frequencies are (in Hz) 225, 369, 476, 680, 800, 1,094, 1,200, 1,504, and 1,628, and the corresponding decrease in intensity is 60 dB in 15 seconds. This gives a rather unnatural "electronic" decay. Second, we hear the different components given different decay times (in general, the higher the component, the shorter the decay), which makes the sound more natural. The decay times in seconds are 15, 9, 4.9, 5.3, 3.75, 3, 2.25, 1.5, and 1.1. Third, we hear two of the components split into pairs of neighboring frequencies (225 is replaced by 224 and 225, and 369 is replaced by 368 and 369.7). The beats add liveliness and warmth to the sound (see Chapter 12).

*Example 3.7. Bell-like sounds followed by fluid textures.*

This example presents a musical fragment with bell-like tones (part A in the left-hand figure on

the facing page), followed by fluid textures (part B). These two parts are closely related. The unequally spaced components of the bell-like tones (in part A) have the same amplitude envelope: a short attack followed by an exponential decay. However, the decays of the various components have different durations; part B is deduced from part A by simply changing the envelope to a smooth bell-like shape. The various components will not reach their maximum amplitude at the same time; hence, instead of fusing into a bell-like attack, they yield textures in which the individual components are dispersed (like white light through a prism) and can thus be more clearly heard separately. This shows that changing one curve (see Chapter 13) can cause subtle transformations of a sound: parts A and B sound very different, yet are closely related because their frequency components (that is, the underlying harmony) are the same.

*Example 3.8. Drumlike sounds.*

This example presents, first, eight percussive sounds similar to the sounds produced by tom-toms, followed by sounds reminiscent of those of snare drums. The tom-tom-like sounds were obtained with a percussive envelope that decayed 48 dB in .2 to .3 second, applied over a wave comprising partials with frequencies such as 200, 320, 440, and 460 Hz. The snarelike effect is obtained by adding a noise band centered at 4,000 Hz, of 3,000-Hz bandwidth, with the same envelope.

Then we hear sounds reminiscent of those made by Indian drums. Again, a percussive envelope (not as sharp as formerly) is applied to a set of inharmonic partials. Some of the sounds have variable pitch, which gives the impression of stretching or unstretching a drum membrane.

*Example 3.9. The electronic violin.*

Brasslike sound produced by Max Mathews's electronic violin when equipped with an electronic filter that makes higher partials more prominent as intensity increases.

■ Bell-like sounds followed by
fluid textures.

■ Transition between two timbres.

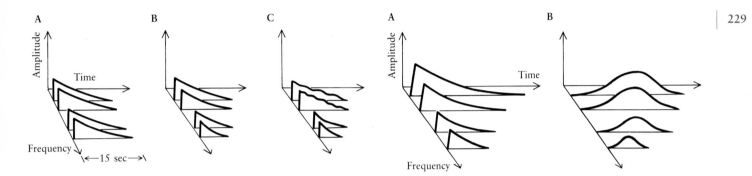

### Example 3.10. Voice synthesis.

The frequency-modulation technique described in Appendix F was used to synthesize these voicelike sounds. As in Example 2.7, we hear the importance of a mixture of periodic and random vibrato in fusing the partials of a synthetic voice. Part *a* goes from the fundamental (400 Hz) to the fundamental and partials, to fundamental and partials all with the same vibrato, which causes them to fuse into a single voicelike sound. Part *b* has three fundamentals—400, 500, and 600 Hz. The partials are subsequently added and then a different rate of vibrato is added to each complex, so that the complex tone fuses into three distinct voices.

### Example 3.11. Effect of envelope on timbre.

How a sound rises and falls affects its timbre. If the intensity rises rapidly and falls slowly, we tend to hear the sound as plucked or struck. Here we compare the effects of a trapezoidal waveform (rise, constant, decay) to those of an abrupt rise followed by an exponential (gradually falling) decay, using two different waveforms. The notes are .25 second in duration. The first waveform is made up of fifteen harmonics of equal amplitude:

a. trapezoidal;
b. exponential;
c. trapezoidal.

The second waveform is a sine wave:

a. trapezoidal;
b. exponential;
c. trapezoidal.

### Example 3.12. Effect of spectrum on timbre.

These synthetic vowels (ah and ee) have both pitch and brightness (an aspect of timbre). The pitch is determined by the fundamental frequency. The brightness (for a given pitch) is determined by whether the formants (ranges of frequency in which the harmonics are intense) are at a lower frequency (ah) or a higher frequency (ee). A vowel can have a higher or lower pitch regardless of brightness. Although brightness may increase with pitch, a sound of a given pitch may be more or less bright. First, we hear the vowel *ah*, two pitches a fifth apart; then the vowel *ee*, two pitches a fifth apart. Second, we hear the vowel *ah*, two pitches a fifth apart sounded together; then the vowel *ee*, two pitches a fifth apart sounded together. With a different vibrato on each frequency, we hear a "chorus" instead of a single voice.

### Example 3.13. Transition between two timbres.

As in Example 3.2, this example uses the effect of envelope on timbre. A steady waveform made up of eleven harmonics is controlled by an amplitude envelope. From the first to the twelfth notes played, this envelope is gradually changed from a smooth attack and short decay (see part A of the above figure) to one with a sharp attack immediately followed by a long decay (part B of the figure). This change in the envelopes causes a transition from a sharp reedy timbre to a plucked-string quality.

### Example 3.14. A musical fragment.

This fragment explores the transformation demonstrated in Example 3.7, as well as another transformation akin to the harmonic dispersion demonstrated in Example 1.6. The inharmonic components are shifted in time while keeping their percussive envelope, so that, instead of fusing into a bell tone, they sound more like many tiny bells.

### Examples for Chapter 14: Paradox and Illusion

### Example 4.1. The octave illusion.

You must use headphones to hear this illusion, illustrated in the figure on page 189. A right-handed listener will hear what is shown in part B of the figure. The right ear disregards the low tones, and the left ear disregards the high tones.

### Example 4.2. The scale illusion.

Again, you must use headphones to hear this illusion, illustrated in the figure on page 190. The

230

left and right ears hear alternate notes of a descending and an ascending scale. The right ear hears a scale that goes down and then up; the left ear hears a scale that goes up and then down.

*Example 4.3. Musical streaming: the Wessel illusion.*

We tend to follow successive sounds of similar timbre and different pitch. These sequences of notes illustrate the effect of similar and different timbres on successive notes (see the figure on page 191). If all the notes are played with a similar timbre, we hear the repeating pattern of three rising notes, just as in part A of the figure.

When two distinctive timbres are used alternately, we hear two patterns of three, more widely spaced descending notes, as shown in part B of the figure on page 191.

*Example 4.4. Two sounds of different pitch.*

Most listeners find the pitch of the first tone higher than that of the second tone (although some find the contrary), and many find the second tone shriller in timbre. The paradoxical fact is that the second tone is deduced from the first by doubling each of its frequencies, as happens when one plays a tape recording at twice the speed. One would expect the pitch to rise by an octave, but instead it goes down by a small amount. When we depart from "normal" sounds, pitch is not always linked with frequency. Here the first tone comprises stretched octaves; the frequency of the $n$th component is given by $f_n = 2f_{n-1}(1 + \alpha)$, with $\alpha \ll 1$ (here $\alpha = 0.03$), and the $f_n$ ranges between 50 and 8,000 Hz. When frequencies are doubled, each component becomes somewhat lower in frequency than the nearest component of the first sound, except at the low and high ends. The perceptual phenomenon calls for a "proximity principle" for tonal perception. The ear hears a small descent more readily than a large ascent and tends to fuse the many-octave components.

*Example 4.5. An endlessly descending tone.*

This example is similar to Shepard's endlessly ascending (or descending) scale. The ten-octave partials that glide down are controlled in amplitude by a spectral envelope that tapers off at low and high frequencies. After a descent of one octave, the components are exactly the same as at the beginning, because the lowest component can be dropped and a higher component can be introduced insidiously; hence, the descent can continue forever. The second channel is shifted in time by 0.2 second relative to the first. This gives a spatial feeling, which disappears if one listens to only one channel.

*Example 4.6. A tone going both up and down.*

The tone presented here glides down the scale (try following it by singing), but at the same time it becomes gradually shriller, and it obviously ends much higher than the initial pitch. While the octaves glide down, the peak of the spectral distribution goes up; that is, the higher components are gradually strengthened, the lower ones weakened.

*Example 4.7. A steady beat speeding up.*

A steady beat is "gradually doubled" in speed by introducing progressively a beat twice as fast, and so on. Then one hears the reverse, a steady fast beat gradually slowing down by demultiplication. The slowing beat has a harplike timbre; it is, in fact, obtained from a single harp tone recorded, digitized, and processed by a computer.

*Example 4.8. A tone that goes both up and down and that both accelerates and slows down while rotating in space.*

In pitch, this example is the inverse of Example 4.6 but is similar rhythmically: the beat constantly speeds up, and at the same time density diminishes. The rhythmical "trick" is schematized in the figure on page 195: we have the effect of an acceleration, but the final beat is much

slower than the initial one. The feeling of rotation uses control both of amplitude in the left and right channels and of the ratio between direct and reverberant sound, as was suggested by Chowning. The sound source seems to move as shown in the figure on the facing page.

*Example 4.9. Chorus and spatial effects.*

When several neighboring frequency components are not equally spaced, their combination produces an irregular amplitude modulation, which can evoke the "chorus" or "choir" effect of several singers or instruments heard in unison. The effect can also be simulated by an appropriate random amplitude modulation of the sound wave. The first part of this example presents a gliding tone that acquires a choruslike quality by means of random amplitude modulation.

The second part of the example shows that beating can induce spatial effects (as also in Example 4.5). Here a sustained sound made up of sine waves of frequencies corresponding to the chord C#, E, A#, F is mixed with a copy of itself delayed by a varying amount of time. This is akin to the Doppler effect; so the beating of the two sounds is interpreted by the ear as though the sound was somehow spread in space, like a sail in the wind.

*Example 4.10. A musical fragment.*

The fragment exploits chiefly the phasing technique. When waves with many partials beat, different partials beat at different rates and come into prominence at different times. By using a defective harmonic series whose frequencies only approximate those of a given chord, the phasing technique can be applied to a set of partials, those included in this chord, rather than to a tone made up of all harmonic partials below some frequency. For instance, at the beginning of this example, the waveform includes only harmonics 9, 14, 16, 22, 28, and 40 of a fundamental at 23.1 Hz, a very low F#. These correspond, with slight pitch deviations, to a chord G#, D, F#, D, E, A#.

■ Apparent movement of
sound in Example 4.8.

*Example 4.11. Variants of a melodic motif.*

This illustrates the role of melodic "contour" as
a significant characteristic of a melody. The
motif is first played, then expanded (or
stretched) pitchwise, while its contour, that is,
the pattern of succession of ups and downs, is
preserved (the rhythmic pattern is somewhat
varied). To the ear, these variants seem related.

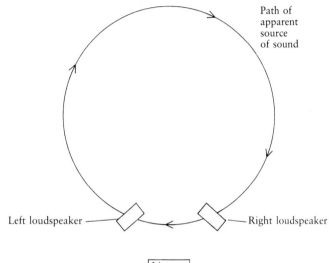

Path of
apparent
source
of sound

Left loudspeaker ─────  ───── Right loudspeaker

Listener

# Sources of Illustrations

pages xiv and 1
Estate of Béla Bartók.

page 3 (left)
Courtesy of the Conservatoire Royal de Musique, Brussels.

page 3 (middle)
Photography by William Gordon Davis/Black Star.

page 3 (right)
Courtesy of King Musical Instruments, Inc., Eastlake, Ohio.

page 4
Courtesy of Lawren Schoenberg and the Arnold Schoenberg Institute.

page 5
Courtesy of Gerald Strang.

page 6 (left)
Music Division, The New York Public Library at Lincoln Center; Astor, Lenox, and Tilden Foundations.

page 6 (right)
Photography by Peter Moore.

page 7 (top)
Courtesy of Max V. Mathews.

page 7 (bottom) and page 8 (left and right)
Music Division, The New York Public Library at Lincoln Center; Astor, Lenox, and Tilden Foundations.

page 9 (top)
Courtesy of Otto C. Luening.

page 9 (middle)
Photography by Ann Holloway.

page 9 (bottom)
Courtesy of Stanford University.

page 10 (top)
Photography by Ralph Fassey.

page 10 (bottom)
Photography by Jean-Pierre Armand/courtesy of the Institut de Recherche et Coordination Acoustique/Musique, Paris.

page 11
Courtesy of Lejarin Hiller.

page 12 (left)
Courtesy of Moog Music, Buffalo, New York.

page 12 (right)
Courtesy of New England Digital Company.

page 13
Courtesy of James A. Moorer.

page 14
© 1963 by Hermann Moeck Verlag, Celle. Used by permission of European American Music Distributors Corporation, sole U.S. agent for Hermann Moeck Verlag.

pages 16 and 17
M. P. Möller, Inc., Hagerstown, Maryland.

page 18
The University Museum, University of Pennsylvania.

page 19 (top)
Photography by Zygmunt Haar/Black Star.

page 19 (bottom)
From *The Science of Sound*, by John Tyndall, Philosophical Library, New York, 1964.

page 21
Adapted from a drawing in *Musical Acoustics: An Introduction*, by Donald E. Hall, Wadsworth, 1980.

page 22
From *Harmonie universelle*, by Marin Mersenne/courtesy of Martinus Nijhoff Publishers.

page 23
From *Anecdotal History of the Science of Sound*, by Dayton C. Miller. Copyright 1935 by Macmillian Publishers Co., Inc., renewed in 1963 by The Cleveland Trust Company, Executor.

page 23 (bottom)
Courtesy of the Metropolitan Museum of Art, Fletcher Fund, 1956.

page 25 (top)
Photography by Peter Simon/Black Star.

page 30 (top)
Photography by Saxon Donnelly/courtesy of the University of California, Berkeley.

page 31 (top)
Photography by Philip L. Molten.

page 33
Courtesy of Julio Prol.

page 34
Courtesy of Steinway & Sons.

pages 38 and 39
By Michael Plass and Scott Kim, © 1983 by Scott Kim.

page 43 (bottom)
Courtesy of Elizabeth A. Cohen, Andrew Schloss, and Eric Schoen.

page 45 (top)
Courtesy of the American Institute of Physics, Niels Bohr Library.

page 45 (bottom)
From *On the Sensations of Tone as a Physiological Basis for the Theory of Music,* by Hermann von Helmholtz.

page 46 (top)
Photography by Barbara Pfeffer/Black Star.

page 46 (bottom)
Photography by Bob Shamis.

page 47 (bottom left)
Photography by Constantine Manos/Magnum.

page 47 (bottom right)
Courtesy of Cliché Publimages. Musée Instrumental du Conservatoire National Supérieur de Musique, Paris.

page 48 (top left)
Photography by William P. Gottlieb.

page 48 (top right)
Photography by Bob Shamis.

page 48 (bottom) and page 49
Courtesy of King Musical Instruments, Inc., Eastland, Ohio.

page 50 (top)
Adapted from a graph in "The Acoustics of Violin Plates," by Carleen Maley Hutchins. Copyright © 1981 by Scientific American, Inc. All rights reserved.

page 51 (top left)
Nicolet Scientific Corporation/courtesy of Max V. Mathews.

page 51
Courtesy of Elizabeth A. Cohen, Scott Forster, Mickey Hart, and The Center for Computer Research in Music and Acoustics at Stanford University.

page 52
By A. M. S. Quinn/courtesy of Max V. Mathews.

page 53 (top)
Courtesy of Elizabeth A. Cohen.

page 55
Courtesy of Elizabeth A. Cohen.

page 57
Courtesy of Elizabeth A. Cohen, The Center for Computer Research in Music and Acoustics at Stanford University, and The Grateful Dead.

pages 60 and 61
Photography by Constantine Manos/Magnum.

page 63
Courtesy of Elizabeth A. Cohen.

pages 72 and 73
© 1963 by Hermann Moeck Verlag, Celle. Used by permission of European American Music Distributors Corporation, sole U.S. agent for Hermann Moeck Verlag.

page 75
Copyright © 1947 by Associated Music Publishers, Inc. All rights reserved. Used by permission.

page 77 and page 79 (top)
Adapted from graphs in *Experiments in Tone Perception,* by R. Plomp, Institute for Perception RVO-TNO, National Defense Research Organization TNO, Soesterberg, The Netherlands, 1966.

page 78
Adapted from a graph in "Critical Band Width in Loudness Summation," by E. Zwicker, E. G. Flottorp, and S. S. Stevens, *The Journal of the Acoustical Society of America* 29(1957):548.

pages 82 and 83
Photography by Martin Dain/Magnum.

page 88 (left and right)
Courtesy of the Institut voor Perceptie Onderzoek, Eindhoven, The Netherlands.

page 92
From *Harmonie universelle,* by Marin Mersenne/courtesy of Martinus Nijhoff Publishers.

pages 94 and 95
Photography by Al Stephenson, © Al Stephenson/Woodfin Camp & Associates.

page 97 (upper left and right)
Adapted from drawings in *Waves and the Ear,* by Willem van Berjeick, John R. Pierce, and Edward E. David, Jr., copyright © 1960 by Anchor Books. Reprinted by permission of Doubleday & Company, Inc.

page 97 (bottom)
Adapted from a drawing in "Neuroanatomy of the Auditory System," by R. R. Gacek in *Foundations of Modern Auditory Theory,* volume 2, J. V. Tobias, editor, Academic Press, 1972.

pages 99 and 100
Adapted from drawings in *Waves and the Ear,* by Willem van Berjeick, John R. Pierce, and Edward E. David, Jr., copyright © 1960 by Anchor Books. Reprinted by permission of Doubleday & Company, Inc.

page 102
From *The Wonders of Acoustics,* by Rodolphe Radau, Scribner's, 1886.

page 103
Photography by Nobu Arakawa/Image Bank.

page 107 (left)
Adapted from a graph in *Modern Sound Reproduction,* by Harry F. Olson, Robert E. Krieger, 1978.

page 111
Adapted from drawings in *Speech and Hearing in Communication,* by Harvey Fletcher, D. van Nostrand, © 1953 D. van Nostrand.

page 113
Reproduced from *The Acoustical Foundations of Music,* by John Backus, by permission of W. W. Norton & Company, Inc. Copyright ©

1969 by W. W. Norton & Company, Inc., and John Murray, Ltd., London.

page 114
Adapted from a graph in "Musical Dynamics," by Blake R. Patterson. Copyright © 1974 by Scientific American, Inc. All rights reserved.

pages 118 and 119
Photography by Jeff Lowenthal, © Jeff Lowenthal/Woodfin Camp & Associates.

page 121
Adapted from graphs in *Speech and Hearing in Communication,* by Harvey Fletcher, D. van Nostrand, © 1953 D. van Nostrand.

page 124
Adapted from graphs in "On the Masking of a Simple Auditory Stimulus," by James P. Egan and Harold W. Hake, *The Journal of the Acoustical Society of America* 22(1950): 622–630.

page 126
Adapted from a graph in "The Acoustics of Singing," by Johann Sundberg. Copyright © 1977 by Scientific American, Inc. All rights reserved.

pages 128 and 129
Courtesy of Manfred R. Schroeder.

page 137
Adapted from an illustration in *Music, Acoustics, and Architecture,* by Leo Beranek, copyright by John Wiley & Sons, Inc.

pages 138 and 139
Courtesy of Elizabeth A. Cohen and the San Francisco War Memorial and Performing Arts Center.

page 140
Adapted from a drawing in *The Collected Papers on Acoustics,* by Wallace Sabine, Harvard University Press, 1922.

page 141
Adapted from a graph in *Architectural Acoustics,* second edition, by K. D. Ginn, Bruel and Kjaer, 1967.

page 142
Photography by James R. Holland/Black Star.

page 143
Photography by Ezra Stoller, © ESTO.

pages 144 and 145
Adapted from graphs in "Acoustical Measurements in Philharmonic Hall," by M. R. Schroeder, B. S. Atal, G. M. Seeler, and J. West, *The Journal of the Acoustical Society of America* 40(1966):434–440.

pages 146 and 147
Photography by Susanne Faulkner Stevens.

page 148 (left and right)
Adapted from graphs in "Acoustical Measurements in Philharmonic Hall," by M. R. Schroeder, B. S. Atal, G. M. Sessler, and J. West, *The Journal of the Acoustical Society of America* 40(1966):434–440.

page 150
Adapted from an illustration in "Construction of a Dummy Head after New Measurements of the Threshold of Hearing," by V. Mellert, *The Journal of the Acoustical Society of America* 51(1960):1359–1361.

page 151 (right)
Courtesy of Manfred Schroeder.

page 152
Photography by Gerard Lobal/courtesy of the Institut de Recherche et Coordination Acoustique/Musique, Paris.

page 153
Adapted from a drawing in "Binaural Dissimilarity and Optimum Ceilings for Concert Halls: More Lateral Sound Diffusion," by M. R. Schroeder, *The Journal of the Acoustical Society of America* 65(1979):958–963.

pages 154 and 155
Courtesy of RCA Corporation.

page 157 (top and bottom)
Adapted from illustrations in "Symposium on Wire Transmission of Symphonic Music and Its Reproduction in Auditory Perspective: System Adaptation," by E. H. Bedell and Iden

Kemey, *Bell System Technical Journal* 13(1934):301–308, © 1934 AT&T.

page 158
Courtesy of DAR Magazine.

page 159
Courtesy of Max V. Mathews.

page 163
Photography by Jacob/Black Star.

page 164
Courtesy of United Press International, Inc.

page 166 (top)
Courtesy of the American Institute of Physics, Niels Bohr Library.

page 166 (bottom)
Courtesy of RCA Corporation.

pages 168 and 169
Photography by Sylvia Johnson, © Sylvia Johnson 1981/Woodfin Camp & Associates.

page 171
Courtesy of Ludwig Industries, Chicago, Illinois.

page 174
Photograph courtesy of Max V. Mathews.

page 175
Adapted from a graph in "Electronic Simulation of Violin Resonances," by M. V. Mathews and J. Kohut, *The Journal of the Acoustical Society of America* 53 (1973): 1620–1626.

page 176
Courtesy of Elizabeth A. Cohen.

page 177
Reproduced with permission from "Phonetics" in *Encyclopaedia Britannica*, fifteenth edition, © 1974 by Encyclopaedia Britannica, Inc.

page 178
Courtesy of Jean-Claude Risset.

page 179
Adapted from a drawing in "Scaling the Musical Timbre," by J. M. Gray, *The Journal of the Acoustical Society of America* 61(1977):1270–1277.

page 180
Photographs courtesy of King Musical Instruments, Inc., Eastlake, Ohio.

pages 181 and 182
Photographs courtesy of Ludwig Industries, Chicago, Ilinois.

page 184
Photography by Michael Samson/courtesy of Amy Malina.

page 185
The Metropolitan Museum of Art, gift of Mrs. Alice Lewisohn Crowley, 1946.

pages 186 and 187
© Scott Kim.

page 194
© BEELDRECHT, Amsterdam/V.A.G.A., New York. Collection Haags Gemeentemuseum— The Hague, 1981.

pages 213 (top and bottom) and 214 (middle)
Adapted from illustrations in *The Technology of Computer Music*, by Max V. Mathews, MIT Press, 1969.

pages 227 and 228
Courtesy of Jean-Claude Risset.

# Index